KEY TO GEOMETRY

Book 8: TABLE OF CONTENTS

TO THE STUDENT:

These books will help you to discover for yourself many important relationships of geometry. Your tools will be the same as those used by the Greek mathematicians more than 2000 years ago. These tools are a **compass** and a **straightedge**. In addition, you will need a **sharpened pencil**. The lessons that follow will help you make drawings from which you may learn the most.

The answer books show **one** way the pages may be completed correctly. It is possible that your work is correct even though it is different. If your answer differs, re-read the instructions to make sure you followed them step by step. If you did, you are probably correct.

Key To Geometry is rooted in the geometry of Euclid. Euclidean geometry is based on five key statements called postulates. Four of Euclid's postulates are short and simple. The third postulate states that one can "draw a straight line from any point to any point." The fifth postulate is longer and more complicated than the others. For two thousand years mathematicians tried to show that the fifth postulate could be rewritten as a combination of the other four. No one ever succeeded. In the early nineteenth century, three mathematicians, a Russian named Lobachevsky, a Hungarian named Bolyai, and a German named Gauss, discovered that changing a few important words in Euclid's fifth postulate would create an entirely new geometry different from Euclid's. Before Lobachevsky, Bolyai and Gauss no one imagined that there could be a non-Euclidean geometry.

On the cover of this booklet Nikolai Lobachevsky works on his new geometry at the University of Kazan in Russia. Geometry was just one of Lobachevsky's varied interests.

® *Key to Fractions, Key to Decimals, Key to Percents, Key to Algebra, Key to Geometry, Key to Measurement,* and *Key to Metric Measurement* are registered trademarks of Key Curriculum Press.
Published by Key Curriculum Press, 1150 65th Street, Emeryville, CA 94608
Printed in the United States of America   11 12 13 LHN 23 22 21 |-78-8

The Perpendicular Bisector of a Segment

Problem: *Construct the perpendicular bisector of a segment.*

A ——————————————————— B

Solution:

1. Draw an arc with center A and a radius more than half of $\overline{AB}$.

2. Draw an arc with center B and the same radius.
   Make the arcs intersect above and below $\overline{AB}$.

3. Label the intersections C and D.

4. Draw line $\overleftrightarrow{CD}$.

5. Label as M the point where $\overleftrightarrow{CD}$ intersects $\overleftrightarrow{AB}$.

6. Compare segments $\overline{AM}$ and $\overline{MB}$.

   Are they congruent? _ _ _ _ _

7. Is M the midpoint of segment $\overline{AB}$? _ _ _ _ _

   Why? _

1. Construct the perpendicular bisector of segment $\overline{RS}$.

Label as B the midpoint of $\overline{RS}$.

2. Construct the perpendicular bisector of the segment below.

The Perpendicular Bisectors of the Sides of a Triangle

1. Construct the perpendicular bisector of side $\overline{AB}$.

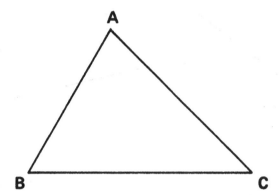

2. Construct the perpendicular bisector of side $\overline{BC}$.

3. Construct the perpendicular bisector of side $\overline{AC}$.

4. Do the perpendicular bisectors intersect in one point? _ _ _ _ _

5. Construct the perpendicular bisector of each side of triangle DEF.

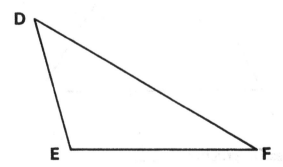

6. Do the perpendicular bisectors intersect in one point? _ _ _ _ _

7. Where do the perpendicular bisectors meet? _ _ _ _ _ _ _ _ _

4

<u>Review</u>

1. Compare the segments.

    Are they congruent? _ _ _ _ _

2. Compare the sides of the triangle.

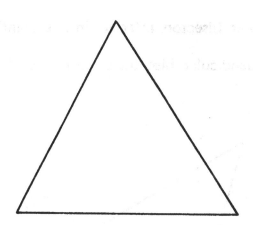

    Are all three sides congruent? _ _ _ _ _

    A triangle whose three sides are congruent is an

    _ _ _ _ _ _ _ _ _ _ _ _ _ triangle.

Problem: *Construct an equilateral triangle.*

A_____B

Solution:

1. Draw an arc with center A and radius $\overline{AB}$.

2. Draw an arc with center B and radius $\overline{AB}$.
   Make the arcs intersect, and label the intersection C.

3. Draw triangle ABC.

4. Construct another equilateral triangle below.

1. Construct the perpendicular bisector of each side of the triangle.

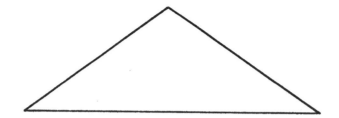

2. Do the perpendicular bisectors intersect in one point? _ _ _ _ _

3. Construct an equilateral triangle.

4. Construct the perpendicular bisector of each side of the equilateral triangle.

5. Do the perpendicular bisectors intersect in one point? _ _ _ _ _

6. The perpendicular bisectors of any triangle _ _ _ _ _ _ _ _ _ _ _ _ intersect in one point.

    (a) always   (b) sometimes   (c) never

7. The perpendicular bisectors _ _ _ _ _ _ _ _ _ _ _ _ _ _ _ intersect inside the triangle.

    (a) always   (b) sometimes   (c) never

For each triangle, <u>predict</u> where you think the perpendicular bisectors will meet.

1. The perpendicular bisectors will meet _ _ _ _ _ _ _ _ _ _ _ _ _ the triangle.

   Construct the perpendicular bisector of each side.

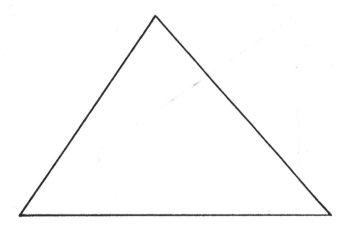

   Was your <u>prediction</u> correct? _ _ _ _ _

2. The perpendicular bisectors will meet _ _ _ _ _ _ _ _ _ _ _ _ _ the triangle.

   Construct the perpendicular bisector of each side.

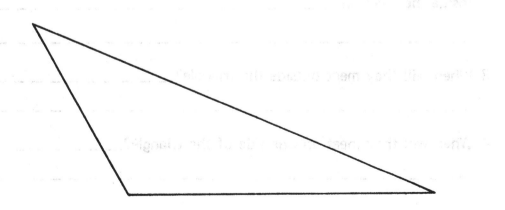

   Was your prediction correct? _ _ _ _ _

1. Predict where you think the perpendicular bisectors will meet.

   The perpendicular bisectors will meet _ _ _ _ _ _ _ _ _ _ _ the triangle.

   Construct the perpendicular bisector of each side.

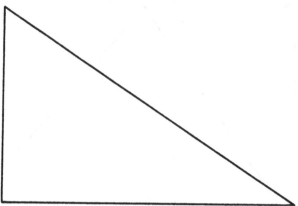

   Was your prediction correct? _ _ _ _ _

2. When will the perpendicular bisectors of the sides of a triangle meet

   inside the triangle? _

   _

3. When will they meet outside the triangle? _ _ _ _ _ _ _ _ _ _ _

   _

4. When will they meet on one side of the triangle? _ _ _ _ _ _ _ _

   _

1. Draw a large triangle and label it MNP.

2. Construct the perpendicular bisector of each side of triangle MNP.

3. Label as Q the point where the perpendicular bisectors intersect.

4. Draw $\overline{QP}$, $\overline{QM}$, and $\overline{QN}$.

5. Compare $\overline{QP}$, $\overline{QM}$, and $\overline{QN}$.

   Are they all congruent? _ _ _ _ _

6. Draw the circle with center Q passing through P.

   Does it pass through M and N? _ _ _ _ _

The Perpendicular Bisectors of the Sides of a Polygon

1. Construct the perpendicular bisector of each side of the quadrilateral.

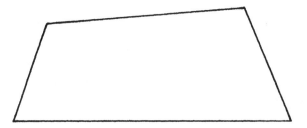

2. Do the perpendicular bisectors intersect in one point? _ _ _ _ _

3. Construct the perpendicular bisector of each side of the quadrilateral below.

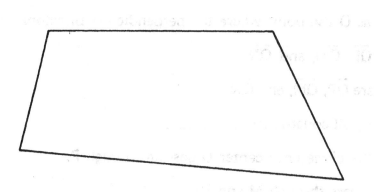

4. Do the perpendicular bisectors intersect in one point? _ _ _ _ _

5. The perpendicular bisectors of the sides of a quadrilateral

_ _ _ _ _ _ _ _ _ _ _ _ _ _ _ intersect in one point.

   (a) always   (b) sometimes   (c) never

1. Choose four points on the circle and label them A, B, C, and D.

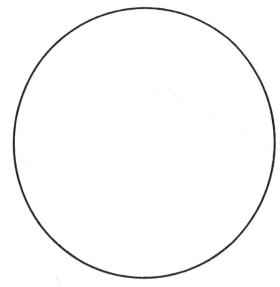

2. Draw quadrilateral ABCD.

3. Construct the perpendicular bisector of each side of ABCD.

4. Do the perpendicular bisectors meet in a point? _ _ _ _ _

5. Draw quadrilateral PQRS.

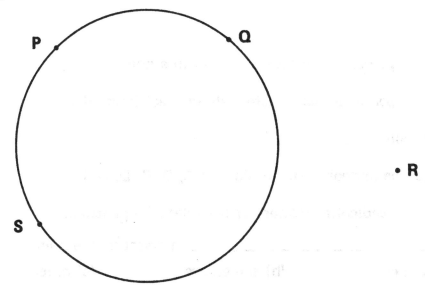

6. Construct the perpendicular bisector of each side of PQRS.

7. Do the perpendicular bisectors meet in a point? _ _ _ _ _

8. When will the perpendicular bisectors of the sides of a quadrilateral

   meet in one point? _

1. Construct the perpendicular bisector of each side of the pentagon.

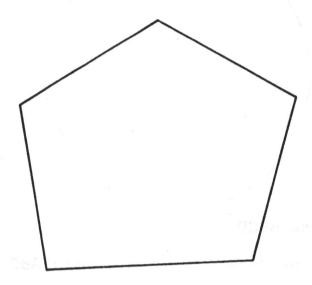

2. Do the perpendicular bisectors meet in a point? _ _ _ _ _

3. Do the perpendicular bisectors all intersect inside the

   pentagon? _ _ _ _ _ _

4. Label the midpoints of the sides as A, B, C, D, and E.

5. The perpendicular bisectors of the sides of a pentagon

   _ _ _ _ _ _ _ _ _ _ _ _ _ _ _ intersect in one point.

   (a) always   (b) sometimes   (c) never

The Medians of a Triangle

1. Find the midpoint of side $\overline{AB}$.

   Label it M.

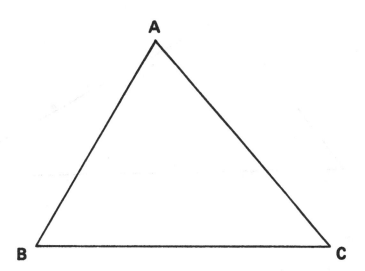

2. Draw $\overline{CM}$.

   The segment joining a vertex and the opposite midpoint is a __median__.
   $\overline{CM}$ is a median of triangle ABC.

3. Find the midpoint of side $\overline{BC}$.

   Label it N.

4. Draw median $\overline{AN}$.

5. Find the midpoint of $\overline{AC}$.

   Label it K.

6. Draw median $\overline{BK}$.

7. Do the medians intersect in one point? _ _ _ _ _

1. Find the midpoint of each side of the triangle.

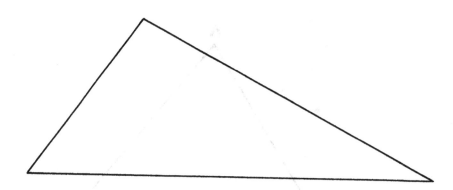

2. Draw the median to each side.

3. Do the medians intersect in one point? _ _ _ _ _

4. Construct an equilateral triangle.

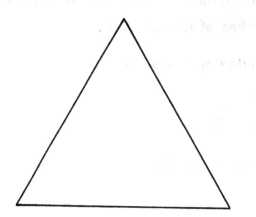

5. Construct the median to each side.

6. Do the medians intersect in one point? _ _ _ _ _

Predict where the medians of each triangle will meet.

1. Will the medians of triangle ABC intersect in one point? _ _ _ _ _

The medians of triangle ABC will intersect _ _ _ _ _ _ _ _ _ the
triangle.

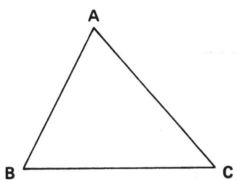

Construct the median to each side.

Were your predictions correct? _ _ _ _ _

2. Will the medians of triangle RST intersect in one point? _ _ _ _ _

The medians will intersect _ _ _ _ _ _ _ _ _ _ the triangle.

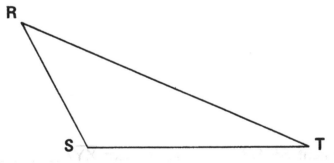

Construct the medians of triangle RST.

Were your predictions correct? _ _ _ _ _

3. The medians of a triangle _ _ _ _ _ _ _ _ _ _ _ _ _ _ intersect
in one point.

(a) always   (b) sometimes   (c) never

4. The medians of a triangle _ _ _ _ _ _ _ _ _ _ _ _ _ _ _ intersect
inside it.

(a) always   (b) sometimes   (c) never

Review

1. Check: Is segment $\overline{AB}$ twice as long as segment $\overline{PQ}$? _ _ _ _ _

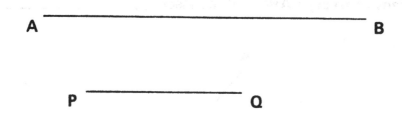

2. Draw a segment three times as long as $\overline{MN}$.

3. Check: Is segment $\overline{RS}$ 1/3 as long as segment $\overline{XY}$? _ _ _ _ _

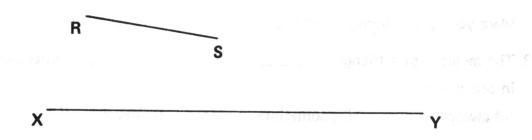

1. Find the midpoint of each side of triangle ABC.

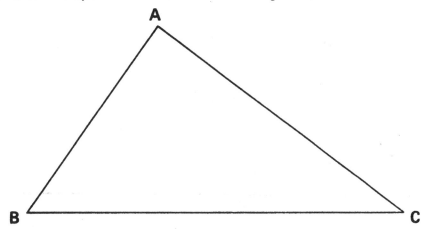

2. Label as M the midpoint of $\overline{AB}$.

   Label as N the midpoint of $\overline{AC}$.

   Label as P the midpoint of $\overline{BC}$.

3. Draw medians $\overline{AP}$, $\overline{BN}$, and $\overline{CM}$.

4. Label as Q the point where they intersect.

5. Compare $\overline{AQ}$ and $\overline{QP}$.

   $\overline{QP}$ is _ _ _ _ _ _ _ _ _ _ _ _ _ _ _ as long as median $\overline{AP}$.

   (a) 1/2   (c) 1/3

   (b) twice   (d) three times

6. Compare $\overline{BQ}$ and $\overline{QN}$.

   $\overline{QN}$ is _ _ _ _ _ _ _ _ _ _ _ _ _ _ _ as long as median $\overline{BN}$.

   (a) 1/2   (c) 1/3

   (b) twice   (d) three times

7. Compare $\overline{CQ}$ and $\overline{QM}$.

   $\overline{QM}$ is _ _ _ _ _ _ _ _ _ _ _ _ _ _ _ as long as median $\overline{CM}$.

   (a) 1/2   (c) 1/3

   (b) twice   (d) three times

The Perpendicular from a Point to a Line

Problem: *Construct the perpendicular from the point to the line.*

P .

A   B

Solution:

1. Draw an arc with center P which intersects line $\overleftrightarrow{AB}$ in two points.

2. Label the points of intersection as R and S.

3. Draw an arc with center R below line $\overleftrightarrow{AB}$.

4. Draw an arc with center S and congruent radius.
   Make the arcs intersect.

5. Label their intersection Q.

6. Draw $\overleftrightarrow{PQ}$.

   $\overleftrightarrow{PQ}$ is _ _ _ _ _ _ _ _ _ _ _ to $\overleftrightarrow{AB}$.

• X

C   D

7. Construct the perpendicular from point X to line $\overleftrightarrow{CD}$.
   Hint: Extend line $\overleftrightarrow{CD}$ to the left.

1. Construct the perpendicular from the point to the line.

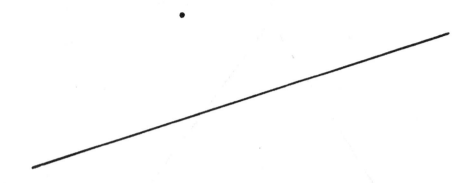

2. Construct the perpendicular from P to line $\overleftrightarrow{RQ}$.

P .

R   Q

3. Construct the perpendicular from vertex A to line $\overleftrightarrow{BC}$.

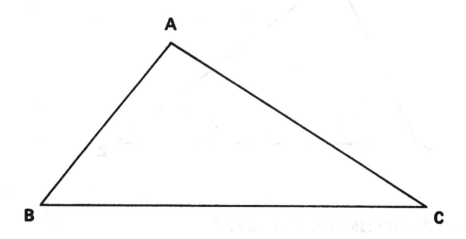

The Altitudes of a Triangle

1. Construct the perpendicular from point A to side $\overline{BC}$.

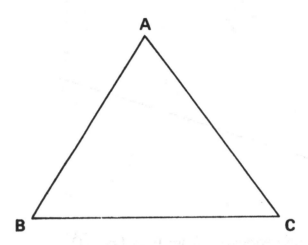

The perpendicular from a vertex to the opposite side of a triangle is an **altitude**.

2. Construct the perpendicular from vertex B to side $\overline{AC}$.

   This line is an _ _ _ _ _ _ _ _ _ _ of triangle ABC.

3. Construct the altitude from X to side $\overline{YZ}$.

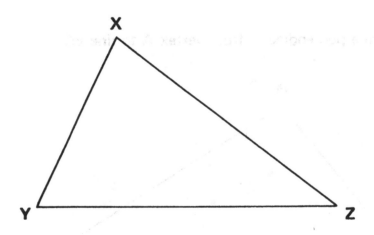

4. Construct the altitude from Y to side $\overline{XZ}$.

   Hint: Extend side $\overline{XZ}$.

1. Construct the altitude from point A to side $\overline{BC}$.

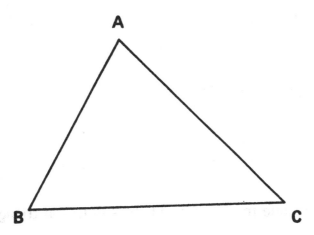

2. Construct the altitude from point B to side $\overline{AC}$.

3. Construct the altitude from point C to side $\overline{AB}$.

4. Do the altitudes intersect in one point? _ _ _ _ _

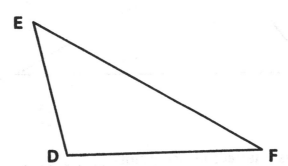

5. Extend side $\overline{DF}$ and $\overline{DE}$.

6. Construct the altitude to each side of the triangle.

7. Extend the altitudes so that they intersect.

8. Do the altitudes intersect in one point? _ _ _ _ _

1. Construct an equilateral triangle.

2. Construct the altitude to each side of the equilateral triangle.

3. Do the altitudes intersect in one point? _ _ _ _ _

4. Construct the altitude to each side of triangle XYZ.

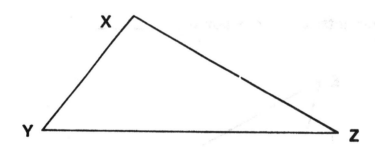

5. Do the altitudes intersect in one point? _ _ _ _ _

6. The altitudes of a triangle _ _ _ _ _ _ _ _ _ _ _ _ _ _ _ _ _
   intersect in one point.

   (a) always   (b) sometimes   (c) never

7. The altitudes of a triangle _ _ _ _ _ _ _ _ _ _ _ _ _ _ _ _
   inside the triangle.

   (a) always   (b) sometimes   (c) never

For each triangle, predict where you think the altitudes will intersect.

1. Will the altitudes meet in one point? _ _ _ _ _

   The altitudes will meet _ _ _ _ _ _ _ _ _ _ _ _ _ _ _ _ _ _ _ the triangle.

Construct the altitude to each side of the triangle.

Were your predictions correct? _ _ _ _ _

2. Will the altitudes of the triangle below meet in one point? _ _ _ _

   The altitudes will meet _ _ _ _ _ _ _ _ _ _ _ _ _ _ _ _ _ _ _ the triangle.

Construct the altitude to each side of the triangle.

Were your predictions correct? _ _ _ _ _

1. Construct the altitude to each side of triangle DEF.

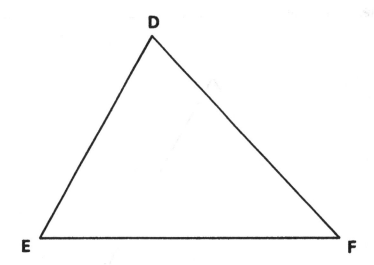

2. Label as Q the point where the altitudes intersect.

3. Draw the circle with center Q passing through D.

4. Does the circle pass through E and F? _ _ _ _ _

Review

1. Construct the perpendicular bisector of each side of the triangle.

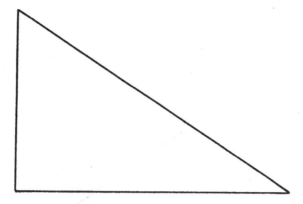

2. Find the midpoint of each side of the triangle.

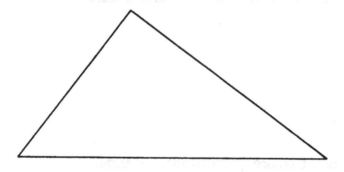

Draw the median to each side of the triangle.

3. Construct the altitude to each side of the triangle.

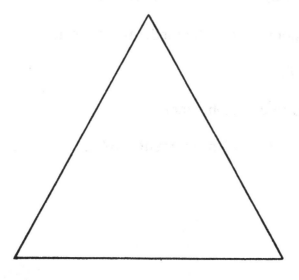

1. Construct the altitude to each side of the triangle.

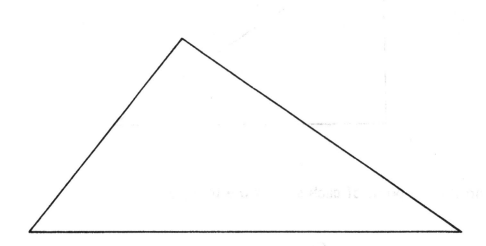

2. Label as A the point where the altitudes intersect.

3. Construct the perpendicular bisector of each side of the triangle.

4. Label as B the point where the perpendicular bisectors intersect.

5. Draw the median to each side of the triangle.

6. Label as C the point where the medians intersect.

7. Draw $\overline{AC}$ and $\overline{CB}$.

8. Are A, B, and C in a straight line? _ _ _ _ _

9. Check: Is segment $\overline{CB}$ 1/3 of segment $\overline{AB}$? _ _ _ _ _

Comparing and Bisecting Angles

Problem: *Compare two angles.*

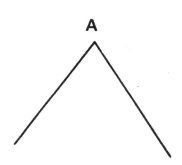
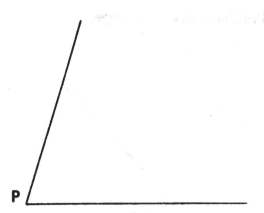

Solution:

1. Draw an arc with center A which intersects both sides of the angle.
   Label as B and C the points where the arc intersects the sides of the angle.

2. Draw an arc with center P and the same radius which intersects both sides of the angle at P.
   Label as Q and R the points of intersection.

3. Compare $\overline{BC}$ and $\overline{QR}$.

   Are they congruent? _ _ _ _ _

4. Are the angles congruent? _ _ _ _ _

5. Compare the angles below.

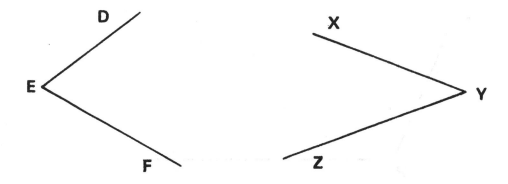

6. Which angle is larger?_ _ _ _ _

Problem: *Bisect an angle.*

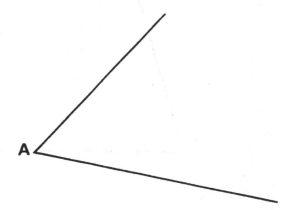

Solution:

1. Draw an arc with center A which intersects both sides of the angle.

2. Label the points of intersection as B and D.

3. Draw an arc with B as center.

4. Use the same radius to draw an arc with D as center. Make the arcs intersect.

5. Label the intersection C.

6. Draw $\overrightarrow{AC}$.

7. Check: Is angle BAC congruent to angle CAD? _ _ _ _ _

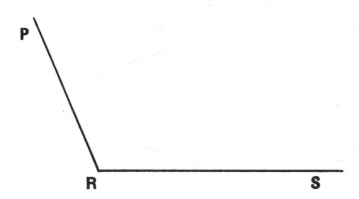

8. Bisect angle PRS.

The Angle Bisectors of a Triangle

1. Bisect angle ABC.

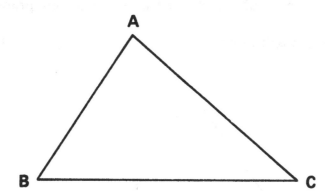

2. Bisect angle ACB.

3. Bisect angle BAC.

4. Do the angle bisectors intersect in one point? _ _ _ _ _

   Do they intersect inside or outside the triangle? _ _ _ _ _

5. Construct an equilateral triangle.

6. Bisect each angle of the equilateral triangle.

7. Do the angle bisectors intersect in one point? _ _ _ _ _

   Do they intersect inside or outside the triangle? _ _ _ _ _

For each triangle, predict where the angle bisectors will intersect.

1. Will the angle bisectors of the triangle intersect in one point? _ _ _ _

   The angle bisectors will intersect _ _ _ _ _ _ _ _ _ _ the triangle.

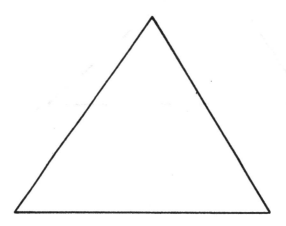

   Construct the bisector of each angle of the triangle.

   Were your predictions correct? _ _ _ _ _

2. Will the angle bisectors of the triangle intersect in one

   point? _ _ _ _ _

   The angle bisectors will intersect _ _ _ _ _ _ _ _ _ _ the triangle.

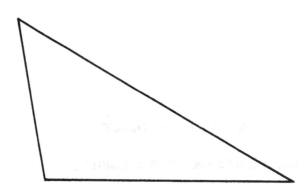

   Construct the bisector of each angle of the triangle.

   Were your predictions correct? _ _ _ _ _

1. Bisect each angle of the triangle.

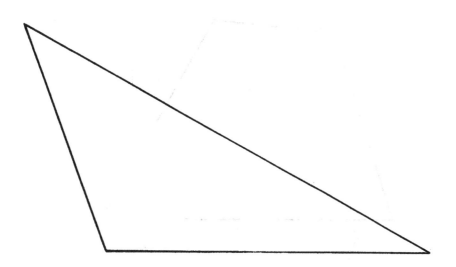

2. Do the bisectors intersect in one point? _ _ _ _ _

3. The bisectors of the angles of a triangle _ _ _ _ _ _ _ _ _ _ _ _ _
   intersect in one point.

   (a) always   (b) sometimes   (c) never

4. The bisectors of the angles of a triangle _ _ _ _ _ _ _ _ _ _ _ _
   intersect inside the triangle.

   (a) always   (b) sometimes   (c) never

1. Bisect each angle of the quadrilateral.

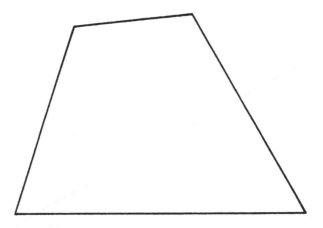

2. Do the bisectors intersect in one point? _ _ _ _ _

3. Bisect each angle of the quadrilateral below.

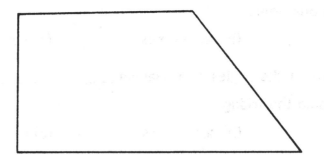

4. Do the bisectors intersect in one point? _ _ _ _ _

5. The angle bisectors of a quadrilateral _ _ _ _ _ _ _ _ _ _ _ _ _ _
   intersect in one point.

   (a) always   (b) sometimes   (c) never

1. Bisect each angle of triangle ABC.

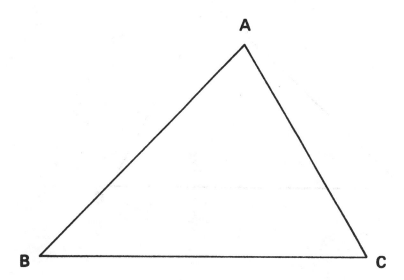

2. Label as P the point where the bisectors intersect.

3. Compare segments $\overline{PA}$, $\overline{PB}$, and $\overline{PC}$.

   Are they congruent? _ _ _ _ _

4. Do you think you can draw a circle with center P which passes

   through A, B, and C? _ _ _ _ _

5. Try to draw a circle with center P which passes through A, B, and
   C.

1. Bisect each angle of triangle ABC.

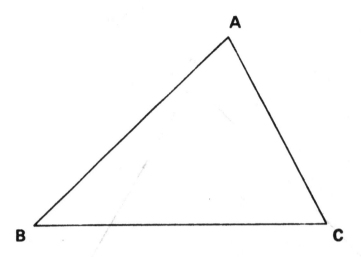

2. Label as P the point where the bisectors intersect.

3. Construct the perpendicular from P to side $\overline{BC}$.
   Label as D the point where the perpendicular from P to side $\overline{BC}$.

4. Construct the perpendicular from P to side $\overline{AB}$.
   Label as E the point where the perpendicular intersects side $\overline{AB}$.

5. Construct the perpendicular from P to side $\overline{AC}$.
   Label as F the point where the perpendicular intersects side $\overline{AC}$.

6. Compare segments $\overline{PD}$, $\overline{PE}$, and $\overline{PF}$.

   Are they all congruent? _ _ _ _ _

7. Can you draw a circle with center P which passes through points D,

   E, and F? _ _ _ _ _

Practice Test

1. Construct the perpendicular bisector of each side.

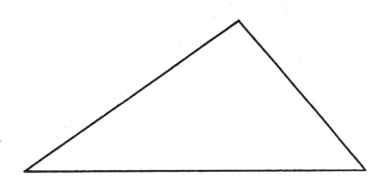

2. Construct the altitude from B to side $\overline{AC}$.

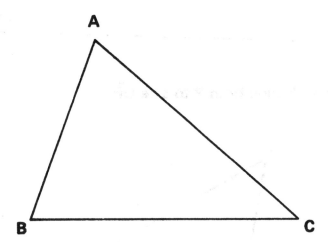

3. Construct the median from B to side $\overline{AC}$.

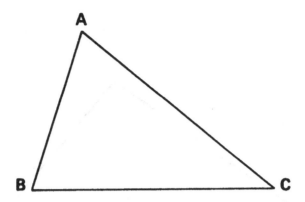

Label the median $\overline{BD}$.

4. Construct the bisector of each angle.

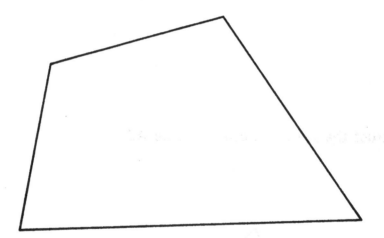

5. Construct the altitude from P to side $\overline{QR}$.

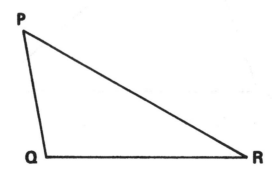

The Perpendicular to a Line through a Point on the Line

Problem: *Construct the perpendicular to a line at a point on the line.*

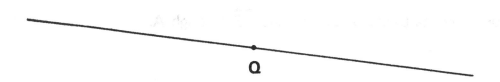

Q

Solution:

1. Draw two arcs with center Q and the same radius which intersect the line in two points.

2. Label the points of intersection A and B.

3. Draw an arc with center A and radius longer than $\overline{AQ}$ above the line.

4. Use the same radius to draw an arc with center B.
   Make the arcs intersect.

5. Label the point of intersection R.

6. Draw the line through Q and R.

7. Is $\overleftrightarrow{QR}$ perpendicular to $\overleftrightarrow{AB}$? _ _ _ _ _

1. Construct the perpendicular to line $\overleftrightarrow{MN}$ at point M.

2. Construct the perpendicular to line $\overleftrightarrow{AB}$ through A.

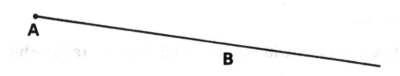

3. Check: Are the lines perpendicular? _ _ _ _ _

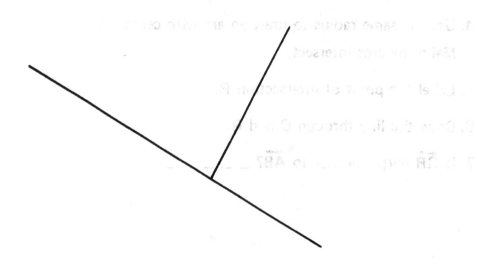

Parallel Lines

1. Construct a line perpendicular to $\overrightarrow{AB}$ through point A.

C   D
_____

_____
        A   B

2. Label as T the point where the perpendicular intersects $\overleftrightarrow{CD}$.

3. Check: Is $\overleftrightarrow{CD}$ perpendicular to $\overrightarrow{AT}$? _ _ _ _ _

4. If two lines are perpendicular to the same line, they are <u>parallel</u>.

   Are lines $\overleftrightarrow{CD}$ and $\overleftrightarrow{AB}$ perpendicular to $\overrightarrow{TA}$? _ _ _ _ _

   Are lines $\overleftrightarrow{CD}$ and $\overleftrightarrow{AB}$ parallel? _ _ _ _ _

5. Construct a line perpendicular to $\overrightarrow{XY}$ through point X.

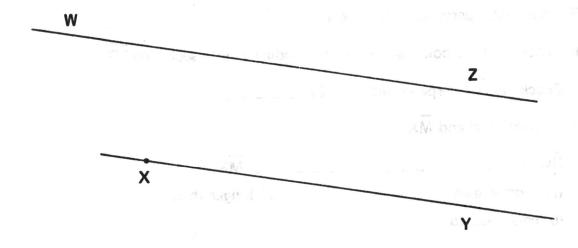

6. Label as P the intersection of the perpendicular and line $\overleftrightarrow{WZ}$.

7. Check: Is $\overleftrightarrow{WZ}$ perpendicular to $\overrightarrow{XY}$? _ _ _ _ _

8. Is $\overleftrightarrow{WZ}$ parallel to $\overrightarrow{XY}$? _ _ _ _ _

1. Construct the perpendicular to $\overleftrightarrow{MN}$ through point M.

Y   Z

M   N

2. Label as X the point where the perpendicular intersects line $\overleftrightarrow{YZ}$.

3. Check: Is $\overleftrightarrow{MX}$ perpendicular to $\overleftrightarrow{YZ}$? _ _ _ _ _

4. Is $\overleftrightarrow{YZ}$ parallel to $\overleftrightarrow{MN}$? _ _ _ _ _

5. Construct a perpendicular to $\overleftrightarrow{MN}$ at N.

6. Label as W the point where the perpendicular intersects line $\overleftrightarrow{YZ}$.

7. Check: Is $\overleftrightarrow{NW}$ perpendicular to $\overleftrightarrow{YZ}$? _ _ _ _ _

8. Compare $\overline{NW}$ and $\overline{MX}$.

$\overline{NW}$ is _ _ _ _ _ _ _ _ _ _ _ _ _ _ _ _ $\overline{MX}$.

(a) shorter than   (c) longer than

(b) congruent to

Lines $\overleftrightarrow{AB}$ and $\overleftrightarrow{DC}$ are parallel.

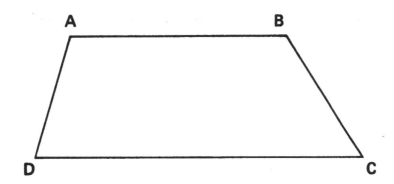

1. Construct the perpendicular from A to $\overleftrightarrow{DC}$.

2. Label as Q the point where the perpendicular intersects $\overleftrightarrow{DC}$.

3. Construct the perpendicular from B to $\overleftrightarrow{DC}$.

4. Label as R the point where the perpendicular intersects $\overleftrightarrow{DC}$.

5. Compare segments $\overline{AQ}$ and $\overline{BR}$.

   $\overline{AQ}$ is _ _ _ _ _ _ _ _ _ _ _ _ _ _ _ $\overline{BR}$.

   (a) shorter than   (c) longer than

   (b) congruent to

1. Choose three points on the bottom line and label them A, B, and C.

R   S

2. Construct the perpendicular to line $\overleftrightarrow{AC}$ through point A. Label as O the point where it intersects line $\overleftrightarrow{RS}$.

3. Check: Is $\overleftrightarrow{AO}$ perpendicular to $\overleftrightarrow{RS}$? _ _ _ _ _

   Are $\overleftrightarrow{AC}$ and $\overleftrightarrow{RS}$ parallel? _ _ _ _ _

4. Construct the perpendicular to line $\overleftrightarrow{AC}$ through point B. Label as P the point where it intersects line $\overleftrightarrow{RS}$.

5. Construct the perpendicular to line $\overleftrightarrow{AC}$ through point C. Label as Q the point where it intersects line $\overleftrightarrow{RS}$.

6. Compare $\overline{AO}$, $\overline{BP}$, and $\overline{CQ}$.

7. The distance between two parallel lines is _ _ _ _ _ _ _ _ _ _ _ _ the same.

   (a) always   (b) sometimes   (c) never

A ———————————————————————————— B

D •———————————————————————————• C

1. Construct the perpendicular to line $\overleftrightarrow{DC}$ at point D.

2. Label as J the point where the perpendicular intersects line $\overleftrightarrow{AB}$.

3. Construct the perpendicular to line $\overleftrightarrow{DC}$ at point C.

4. Label as K the point where the perpendicular intersects line $\overleftrightarrow{AB}$.

5. Compare segments $\overline{DJ}$ and $\overline{CK}$.

   Are they congruent? _ _ _ _ _

6. Do you think the lines are parallel? _ _ _ _ _

   Why? _

Constructing an Angle Congruent to a Given Angle

Problem: *Construct an angle congruent to a given angle.*

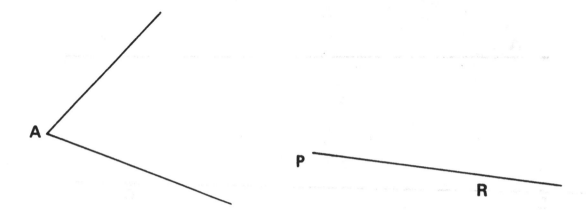

Solution:

1. Draw an arc with center A which intersects both sides of the given angle.

   Label the points of intersection B and C.

2. Use the same radius to draw an arc with center P which intersects ray $\overrightarrow{PR}$.

   Label the point of intersection D.

3. Draw an arc with center D and radius congruent to $\overline{BC}$.

   Make it intersect the arc with center P.

4. Label the point of intersection E.

5. Draw angle EPD.

1. Construct an angle congruent to angle GKL.

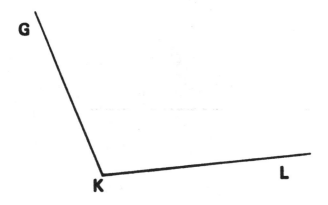

2. Check: Are the angles congruent? _ _ _ _ _

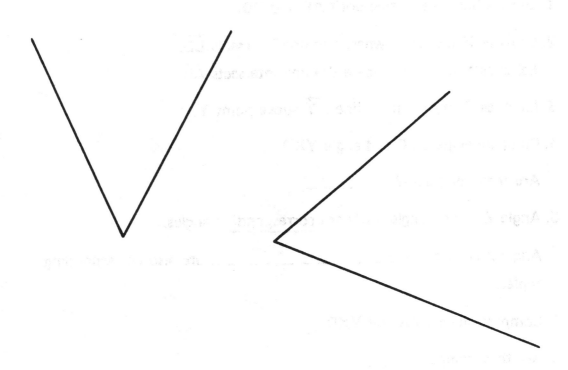

Corresponding Angles

Lines $\overleftrightarrow{AB}$ and $\overleftrightarrow{CD}$ are parallel.

A ————————————————————————— B

C ————————————————————————— D

1. Draw a line intersecting both $\overleftrightarrow{AB}$ and $\overleftrightarrow{CD}$.

2. Label as X the point where the line intersects $\overleftrightarrow{CD}$.
   Label as Y the point where the line intersects $\overleftrightarrow{AB}$.

3. Label as Z any point on line $\overrightarrow{XY}$ above point Y.

4. Compare angle ZYB and angle YXD.

   Are they congruent? _ _ _ _ _

5. Angle ZYB and angle YXD are <u>corresponding</u> angles.

   Angle ZYA and angle _ _ _ _ _ _ _ _ _ _ are also corresponding angles.

6. Compare angle ZYA and YXC.

   Are they congruent? _ _ _ _ _

Lines $\overleftrightarrow{MP}$ and $\overleftrightarrow{RQ}$ are parallel.

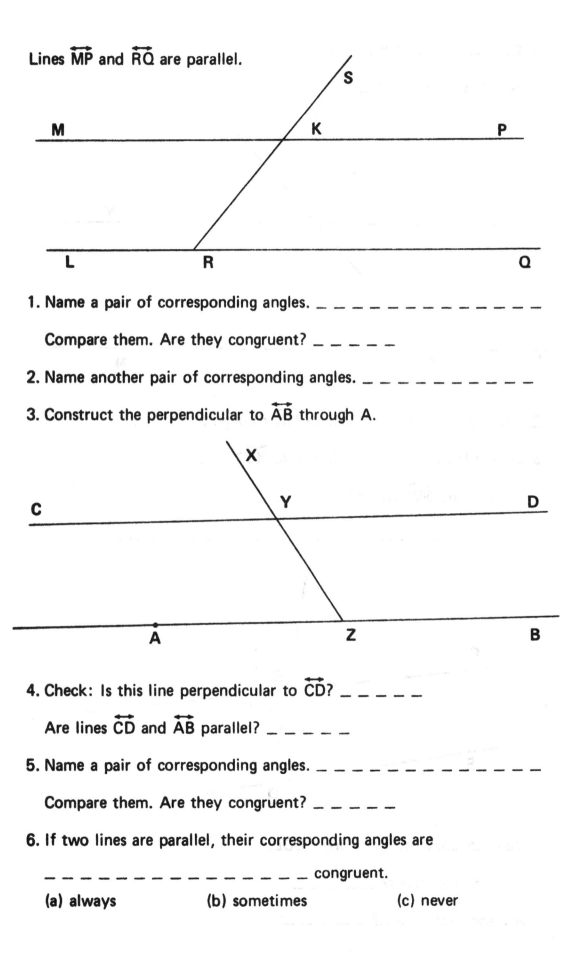

1. Name a pair of corresponding angles. _ _ _ _ _ _ _ _ _ _ _ _ _

   Compare them. Are they congruent? _ _ _ _ _

2. Name another pair of corresponding angles. _ _ _ _ _ _ _ _ _ _ _

3. Construct the perpendicular to $\overleftrightarrow{AB}$ through A.

4. Check: Is this line perpendicular to $\overleftrightarrow{CD}$? _ _ _ _ _ _

   Are lines $\overleftrightarrow{CD}$ and $\overleftrightarrow{AB}$ parallel? _ _ _ _ _

5. Name a pair of corresponding angles. _ _ _ _ _ _ _ _ _ _ _ _ _

   Compare them. Are they congruent? _ _ _ _ _

6. If two lines are parallel, their corresponding angles are

   _ _ _ _ _ _ _ _ _ _ _ _ _ _ _ _ _ congruent.

   (a) always   (b) sometimes   (c) never

1. Compare angle ZXY and angle XMN.

   Are they congruent? _ _ _ _ _

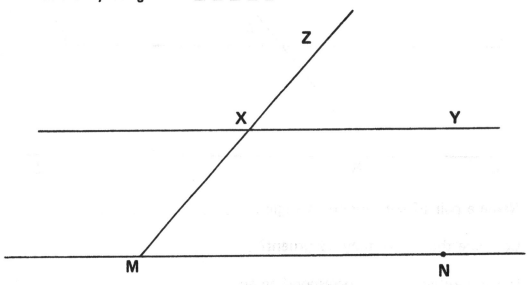

2. Construct the perpendicular to $\overleftrightarrow{MN}$ through N.

3. Check: Is this line perpendicular to $\overleftrightarrow{XY}$? _ _ _ _ _

4. Are $\overleftrightarrow{XY}$ and $\overleftrightarrow{MN}$ parallel? _ _ _ _ _

   Why? _

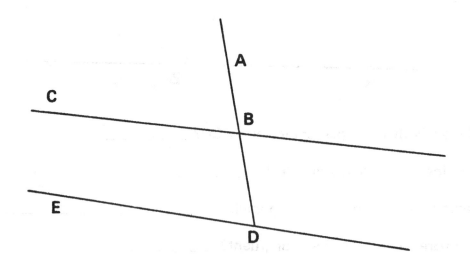

5. Compare angle ABC and angle BDE.

   Are they congruent? _ _ _ _ _

6. Are the lines parallel? _ _ _ _ _

Problem: *To construct a line parallel to $\overleftrightarrow{AB}$.*

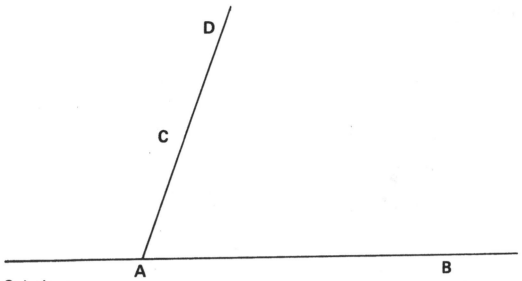

Solution:

1. Draw an arc with center A which intersects both sides of angle CAB. Label the points of intersection P and Q.

2. Use the same radius to draw an arc with center C which intersects $\overrightarrow{CD}$ above C.

   Label the point of intersection R.

3. Draw an arc with center R and radius $\overline{PQ}$ to the right of point C. Make it intersect the arc with center C.

4. Label the point of intersection T.

5. Draw the line through C and T.

6. Do you think $\overleftrightarrow{CT}$ is parallel to line $\overleftrightarrow{AB}$? _ _ _ _ _

7. Construct the perpendicular to line $\overleftrightarrow{AB}$ at point B.

8. Label as S the point where the perpendicular intersects $\overleftrightarrow{CT}$.

9. Check: Is $\overleftrightarrow{BS}$ perpendicular to $\overleftrightarrow{CT}$? _ _ _ _ _

10. Is $\overleftrightarrow{CT}$ parallel to $\overleftrightarrow{AB}$? _ _ _ _ _

   Why? _

Constructing a Line Parallel to a Given Line

Problem: *Construct the line parallel to the given line through the given point.*

P
·

_____
A   B

Solution:

1. Draw the line through A and P.

   Label as M a point on the line above P.

2. At point P construct an angle congruent to angle PAB with $\overrightarrow{PM}$ as one side, and the other side to the right of P.

3. Label the angle MPQ.

4. Is angle MPQ congruent to angle PAB? _ _ _ _ _

5. Is line $\overleftrightarrow{PQ}$ parallel to $\overrightarrow{AB}$? _ _ _ _ _

1. Construct the line parallel to the given line through the given point.

2. Line $\overleftrightarrow{GB}$ is parallel to line $\overleftrightarrow{ED}$.

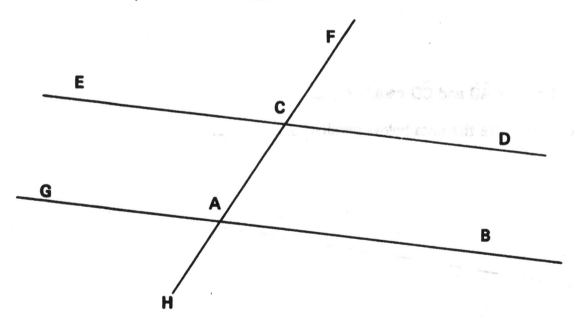

Angle FCD and angle _ _ _ _ _ _ _ _ _ _ _ are corresponding angles.

Angle ECF and angle _ _ _ _ _ _ _ _ _ _ _ are corresponding angles.

Angle ECA and angle _ _ _ _ _ _ _ _ _ _ _ are corresponding angles.

1. Check: Are angles EAB and ACD congruent? _ _ _ _ _

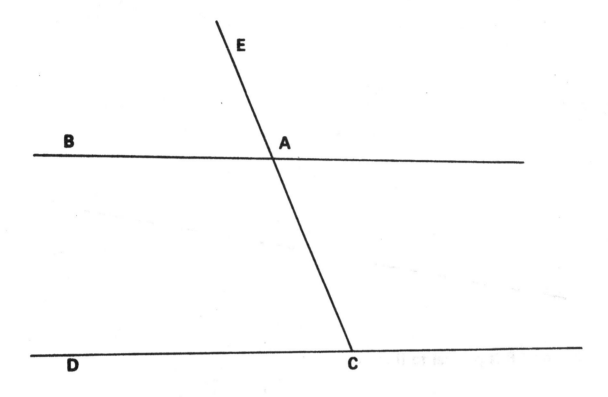

Are lines $\overleftrightarrow{AB}$ and $\overleftrightarrow{CD}$ parallel? _ _ _ _ _

2. Check: Are the lines below parallel? _ _ _ _ _

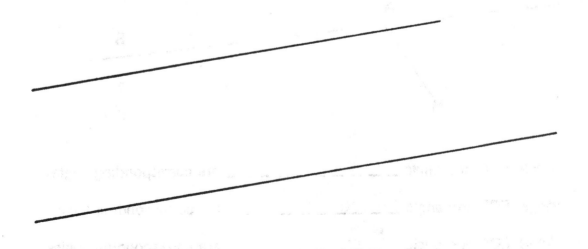

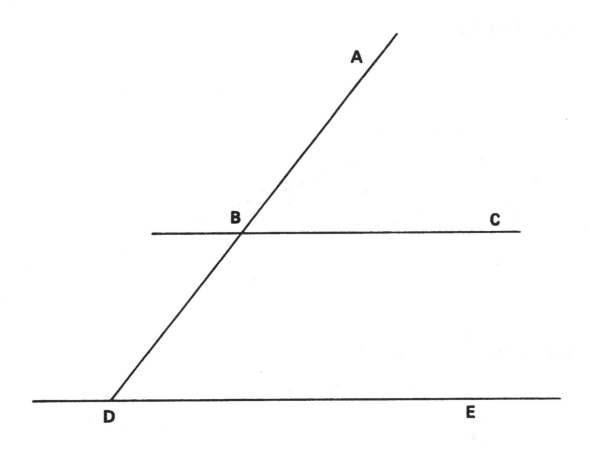

1. Check: Are lines $\overleftrightarrow{BC}$ and $\overleftrightarrow{DE}$ parallel? _ _ _ _ _

   Why?_ _

2. Bisect angle ABC.
   Label the bisector $\overrightarrow{BF}$.

3. Bisect angle BDE.
   Label the bisector $\overrightarrow{DG}$.

4. Check: Are lines $\overleftrightarrow{BF}$ and $\overleftrightarrow{DG}$ parallel? _ _ _ _ _

Review

1. Bisect the segment.

2. Bisect the angle.

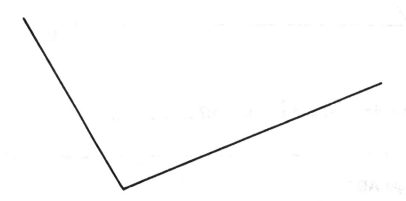

3. Construct a perpendicular to the given line through the given point.

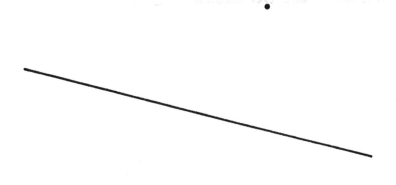

The Line Joining the Midpoints of Two Sides of a Triangle

1. Bisect side $\overline{AB}$.

   Label the midpoint M.

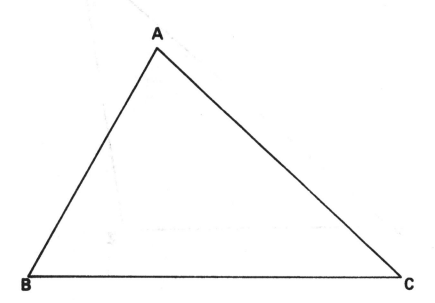

2. Bisect side $\overline{AC}$.

   Label the midpoint N.

3. Draw $\overline{MN}$.

4. Compare angle AMN and angle MBC.

   Are they congruent? _ _ _ _ _

5. Is $\overleftrightarrow{MN}$ parallel to $\overleftrightarrow{BC}$? _ _ _ _ _

   Why? _

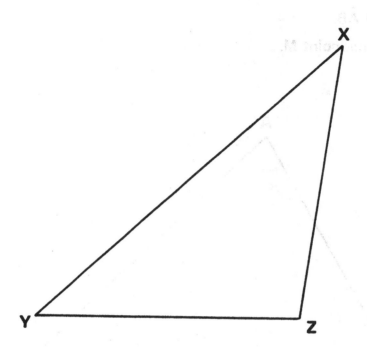

1. Bisect side $\overline{YZ}$.

   Label the midpoint M.

2. Bisect $\overline{XZ}$.

   Label the midpoint N.

3. Draw $\overline{MN}$.

4. Check: Is $\overleftrightarrow{MN}$ parallel to $\overleftrightarrow{XY}$? _ _ _ _ _

5. The line through the midpoints of two sides of a triangle is

   _ _ _ _ _ _ _ _ _ _ _ _ _ _ _ parallel to the third side.

   (a) always   (b) sometimes   (c) never

1. Draw a large triangle and label it DEF.

2. Bisect $\overline{DE}$ and label the midpoint X.

3. Bisect $\overline{DF}$ and label the midpoint Y.

4. Draw $\overline{XY}$.

5. Bisect $\overline{EF}$ and label the midpoint Z.

6. Compare segments $\overline{XY}$ and $\overline{EZ}$ and $\overline{ZF}$.

   Are they all congruent? _ _ _ _ _

7. Segment $\overline{XY}$ is _ _ _ _ _ _ _ _ _ _ _ _ _ _ _ _ _ _ as long as
   segment $\overline{EF}$.

   (a) twice   (c) one-third

   (b) one-half   (d) two-thirds

1. Draw a large triangle and label it ABC.

2. Bisect $\overline{AB}$ and label the midpoint P.

3. Bisect $\overline{AC}$ and label the midpoint R.

4. Draw $\overline{PR}$.

5. Bisect $\overline{BC}$ and label the midpoint T.

6. Draw $\overline{RT}$.

7. Name a parallelogram in the figure. _ _ _ _ _

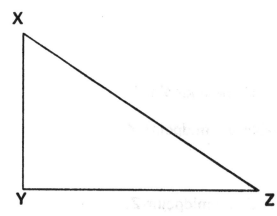

8. Bisect $\overline{XY}$ and label the midpoint M.

9. Bisect $\overline{XZ}$ and label the midpoint Q.

10. Bisect $\overline{YZ}$ and label the midpoint S.

11. Draw $\overline{MQ}$ and $\overline{QS}$.

12. MQSY is called a _ _ _ _ _ _ _ _ _ _ _ _ _ _ .

The Midpoints of the Sides of a Quadrilateral

1. Bisect sides $\overline{AB}$ and $\overline{BC}$ of the quadrilateral.

   Label the midpoints M and N.

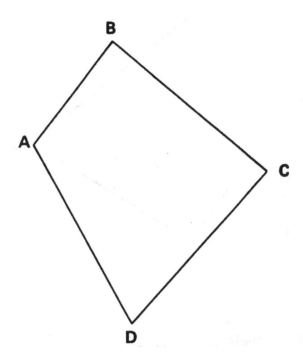

2. Draw $\overline{MN}$.

3. Bisect sides $\overline{AD}$ and $\overline{CD}$.

   Label the midpoints P and Q.

4. Draw $\overline{PQ}$.

5. Are $\overleftrightarrow{MN}$ and $\overleftrightarrow{PQ}$ parallel? _ _ _ _ _

6. Compare segments $\overline{MN}$ and $\overline{PQ}$.

   Are they congruent? _ _ _ _ _ _

1. Bisect sides $\overline{AB}$ and $\overline{BC}$ of the quadrilateral.
   Label the midpoints M and N.

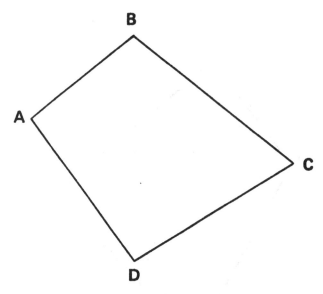

2. Draw $\overline{MN}$ and $\overline{AC}$.

3. Are $\overleftrightarrow{MN}$ and $\overleftrightarrow{AC}$ parallel? _ _ _ _ _

4. Compare segment $\overline{MN}$ and segment $\overline{AC}$.

   Segment $\overline{MN}$ is _ _ _ _ _ _ _ _ _ _ _ _ _ _ _ _ _ segment $\overline{AC}$.
   (a) congruent to   (b) half as long as   (c) one-third as long as

5. Bisect $\overline{AD}$ and $\overline{CD}$. Label the midpoints R and S.

6. Draw $\overline{RS}$.

7. Are $\overleftrightarrow{RS}$ and $\overleftrightarrow{AC}$ parallel? _ _ _ _ _

8. Compare segment $\overline{RS}$ and segment $\overline{AC}$.

   Segment $\overline{RS}$ is _ _ _ _ _ _ _ _ _ _ _ _ _ _ _ _ segment $\overline{AC}$.
   (a) congruent to   (c) one-third as long as
   (b) half as long as

9. Are $\overleftrightarrow{RS}$ and $\overleftrightarrow{MN}$ parallel? _ _ _ _ _

   Are $\overline{RS}$ and $\overline{MN}$ congruent? _ _ _ _ _

1. Bisect side $\overline{PR}$ and label the midpoint A.

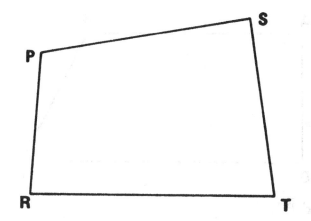

2. Bisect side $\overline{PS}$ and label the midpoint B.

   Bisect side $\overline{ST}$ and label the midpoint C.

   Bisect side $\overline{RT}$ and label the midpoint D.

3. Draw ABCD.

   ABCD is a _ _ _ _ _ _ _ _ _ _ _ _ _ _ _ _ _ .

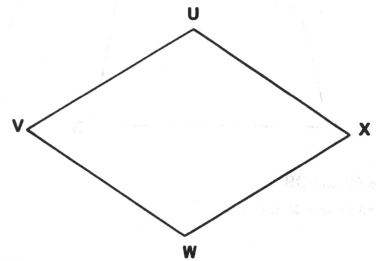

4. Bisect $\overline{UV}$ and label the midpoint E.
   Bisect $\overline{UX}$ and label the midpoint F.
   Bisect $\overline{XW}$ and label the midpoint G.
   Bisect $\overline{VW}$ and label the midpoint H.

5. Draw EFGH.

   EFGH is a _ _ _ _ _ _ _ _ _ _ _ _ _ _ _ _ .

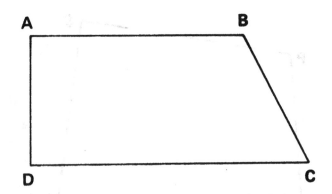

1. Bisect $\overline{AD}$ and $\overline{BC}$.

   Label the midpoints M and N.

2. Draw $\overline{MN}$.

3. Check: Is $\overleftrightarrow{MN}$ parallel to $\overleftrightarrow{DC}$? _ _ _ _ _

   Check: Is $\overleftrightarrow{MN}$ parallel to $\overleftrightarrow{AB}$? _ _ _ _ _

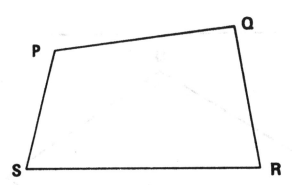

4. Bisect sides $\overline{PS}$ and $\overline{QR}$.

   Label the midpoints X and Y.

5. Draw $\overline{XY}$.

6. Is $\overleftrightarrow{XY}$ parallel to $\overleftrightarrow{SR}$? _ _ _ _ _

   Is $\overleftrightarrow{XY}$ parallel to $\overleftrightarrow{PQ}$? _ _ _ _ _

Review

1. Construct a line parallel to the given line through the given point.

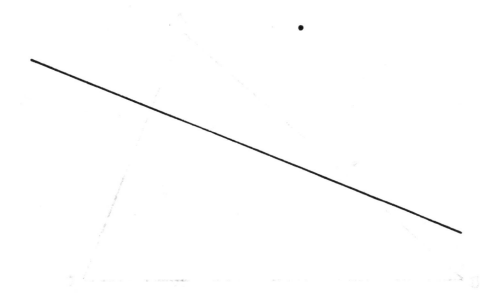

2. Check: Is M the midpoint of segment $\overline{PQ}$? _ _ _ _ _

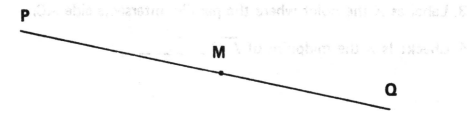

The Line through the Midpoint of One Side of a Triangle and Parallel to Another Side

1. Bisect side $\overline{AB}$.

   Label the midpoint M.

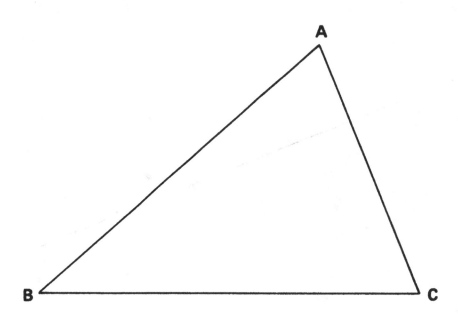

2. Construct the line parallel to $\overleftrightarrow{BC}$ through point M.

3. Label as X the point where the parallel intersects side AC.

4. Check: Is X the midpoint of $\overline{AC}$? _ _ _ _ _

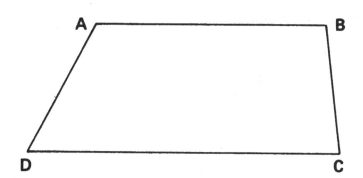

1. Bisect $\overline{AD}$.

   Label the midpoint M.

2. Construct a line parallel to $\overleftrightarrow{DC}$ through point M.

3. Label as X the point where the parallel intersects side $\overline{BC}$.

4. Check: Is X the midpoint of $\overline{BC}$? _ _ _ _ _

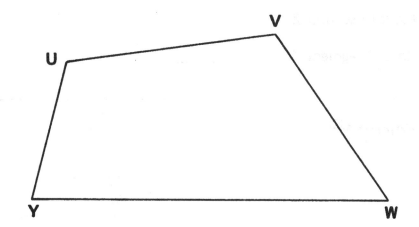

5. Bisect $\overline{UY}$.

   Label the midpoint K.

6. Construct a line parallel to $\overleftrightarrow{YW}$ through point K.

7. Label as L the point where the parallel intersects $\overline{VW}$.

8. Check: Is L the midpoint of $\overline{VW}$? _ _ _ _ _

Dividing a Segment into Congruent Parts

Problem: *Double a given segment.*

A   B

X ————————————————————————

Solution:

1. Draw an arc with center X and radius congruent to $\overline{AB}$ which intersects the ray.

2. Label the intersection Y.

3. Draw an arc with center Y and radius congruent to $\overline{AB}$ which intersects the ray to the right of Y.

4. Label the intersection Z.

5. Is $\overline{XZ}$ double segment $\overline{AB}$? _ _ _ _ _

   Why? _

6. Triple segment $\overline{MN}$.

M
  N

1. Divide segment $\overline{PQ}$ into two congruent parts.

2. Divide segment $\overline{PQ}$ into four congruent parts.

3. Draw a segment below which is 3/4 as long as $\overline{PQ}$.

Review

1. Construct an angle congruent to the given angle with ray $\overrightarrow{AB}$ as one side.

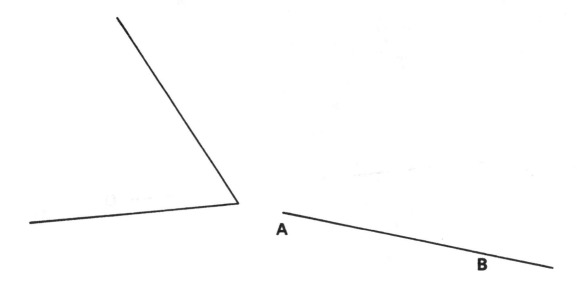

2. Construct a line parallel to the given line through the given point.

3. Bisect the given angle.

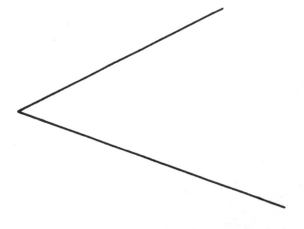

Dividing a Segment into Three Congruent Parts

Problem: *Divide segment $\overline{AB}$ into 3 congruent parts.*

X .

———————————————————————
A   B

Solution:

1. Draw a ray with endpoint A through point X.

2. Draw an arc with center X and radius congruent to $\overline{AX}$ which intersects the ray to the right of X.
   Label the intersection Y.

3. Draw an arc with center Y and radius congruent to $\overline{AX}$ which intersects the ray to the right of Y.
   Label the intersection Z.

4. Are $\overline{AX}$, $\overline{XY}$, and $\overline{YZ}$ all congruent? _ _ _ _ _

5. Draw $\overleftrightarrow{ZB}$.

6. Construct an angle at Y with $\overrightarrow{YA}$ as one side which is congruent to angle AZB.
   Make its other side intersect $\overline{AB}$.
   Label the intersection N.

7. Construct an angle at X with $\overrightarrow{XA}$ as one side which is congruent to angle AZB.
   Make its other side intersect $\overline{AB}$.
   Label the intersection R.

8. Is $\overleftrightarrow{YN}$ parallel to $\overleftrightarrow{ZB}$? _ _ _ _ _

   Is $\overleftrightarrow{XR}$ parallel to $\overleftrightarrow{ZB}$? _ _ _ _ _

9. Are $\overline{AR}$, $\overline{RN}$, and $\overline{NB}$ all congruent? _ _ _ _ _

   Segment $\overline{AB}$ is divided into _ _ _ _ _ _ _ _ _ **congruent parts.**

Problem: *Divide segment $\overline{AB}$ into 3 congruent parts.*

A ———————————————————————————— B

Solution:

1. Draw a ray with endpoint A above $\overline{AB}$.

2. On the ray, lay off 3 congruent segments.
   Label the endpoints X, Y, and Z, in that order.

3. Draw $\overleftrightarrow{ZB}$.

4. Construct the line parallel to $\overleftrightarrow{ZB}$ through Y.
   Label as N the point where it intersects $\overline{AB}$.

5. Construct the parallel to $\overleftrightarrow{ZB}$ through X.
   Label as R the point where it intersects $\overline{AB}$.

6. Check: Are $\overline{AR}$, $\overline{RN}$, and $\overline{NB}$ all congruent? _ _ _ _ _

   Segment $\overline{AB}$ is divided into _ _ _ _ _ _ _ _ _ _ _ congruent parts.

7. Angle XYN and angle YZB are called _ _ _ _ _ _ _ _ _ _ _ angles.

   Angle AXM and angle XYN are called _ _ _ _ _ _ _ _ _ _ _ angles.

1. Divide segment $\overline{AB}$ into three congruent parts.

2. Draw a segment which is 2/3 as long as segment $\overline{AB}$.
   Label its endpoints X and Y.

1. Divide $\overline{FG}$ into three congruent parts.

2. Divide $\overline{RS}$ into five congruent parts.

3. Construct a segment which is 3/5 as long as $\overline{RS}$.
   Label its endpoints A and B.

1. Divide segment $\overline{AB}$ into three congruent parts.

2. Draw a segment which is 1-2/3 as long as segment $\overline{AB}$.
   Label its endpoints R and S.

3. Draw a segment which is 1/6 as long as segment $\overline{AB}$.
   Label its endpoints P and Q.

Dividing an Angle into Congruent Parts

1. Divide the angle into two congruent parts.

    Hint: Bisect the angle.

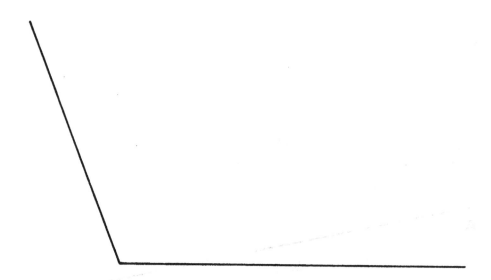

2. Divide the angle into four congruent parts.

3. Construct an angle which is 3/4 as large as the given angle.

Problem: *Double a given angle.*

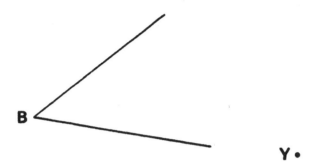

Solution:

1. Draw a ray with endpoint Y to the right of Y.

2. Draw an arc with center B which intersects both sides of the angle at B.

3. Label the points of intersection A and C.

4. With center Y and the same radius, draw a large arc.
   Make the arc intersect the ray, and label the intersection Z.

5. With Z as center and radius congruent to $\overline{AC}$, draw an arc intersecting the large arc.
   Label the intersection V.

6. With V as center and radius congruent to $\overline{AC}$, draw another arc intersecting the large arc.
   Label the intersection X.

7. Draw angle XYZ.

8. Is angle XYZ double angle ABC? _ _ _ _ _

   Why? _

1. Double angle DEF.

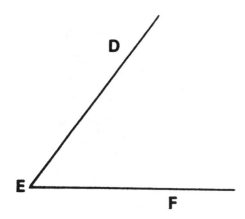

2. Triple angle PQR.

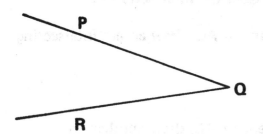

1. Divide angle ABC into two congruent parts.

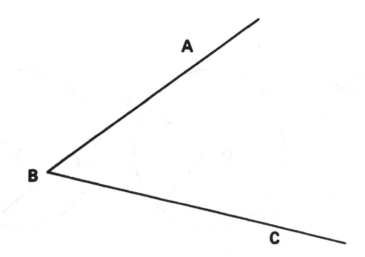

2. Construct an angle 1-1/2 times as large as angle ABC.

Central and Inscribed Angles

1. A <u>central angle</u> is an angle whose vertex is at the center of a circle.

   Which angle below is a central angle? _ _ _ _ _

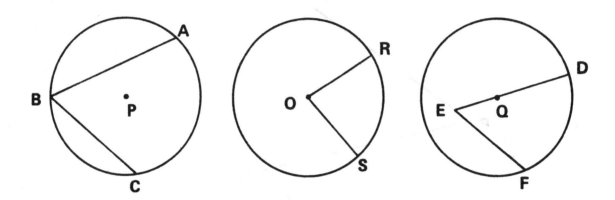

2. An <u>inscribed angle</u> is an angle whose vertex is on the circle.

   Which angle above is an inscribed angle? _ _ _ _ _

3. Draw an inscribed angle in circle X.

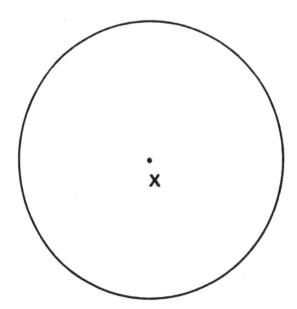

4. Draw a central angle in circle X.

O is the center of the circle.

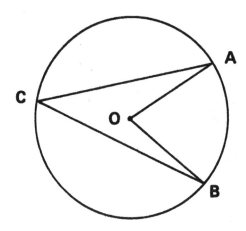

X •

1. Angle _ _ _ _ _ is a central angle.

2. Angle _ _ _ _ _ is an inscribed angle.

3. Construct an angle double angle ACB at X.

4. Label as angle YXZ the angle which is double angle ABC.

5. Compare angle AOB and angle YXZ.

   Are they congruent? _ _ _ _ _

6. The central angle is _ _ _ _ _ times as large as the inscribed angle.

Q is the center of the circle.

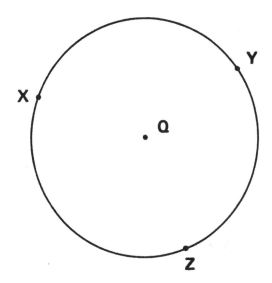

1. Draw a central angle which intersects the circle at points X and Y.

2. Draw an inscribed angle with vertex Z whose sides intersect the circle at points X and Y.

3. Construct an angle double the inscribed angle XZY.

4. Check: Is central angle XQY double inscribed angle XZY? _ _ _ _ _

5. Draw another inscribed angle whose sides intersect the circle at points X and Y. Label its vertex W.

6. Check: Is central angle XQY double inscribed angle XWY? _ _ _ _ _

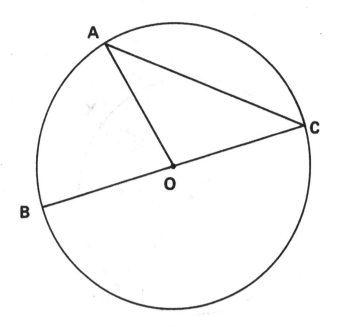

O is the center of the circle.

1. What kind of angle is angle AOB? _ _ _ _ _ _ _ _ _ _ _ _ _ _ _

2. What kind of angle is angle ACB? _ _ _ _ _ _ _ _ _ _ _ _ _ _

3. Bisect angle AOB.

4. Label as X the point where the bisector intersects the circle.

5. Compare angle XOB and angle ACB.

Angle XOB is _ _ _ _ _ _ _ _ _ _ _ _ _ _ _ _ angle ACB.

(a) smaller than   (c) larger than

(b) congruent to

6. Angle ACB is _ _ _ _ _ _ _ _ _ _ _ _ _ _ _ angle AOB.

(a) twice as large as

(b) congruent to

(c) half as large as

7. Line $\overleftrightarrow{XO}$ is _ _ _ _ _ _ _ _ _ _ _ _ _ _ $\overrightarrow{AC}$.

(a) a bisector of   (c) perpendicular to

(b) parallel to

O is the center of the circle.

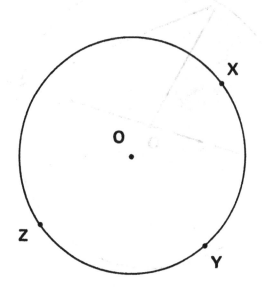

1. Draw central angle XOY.

2. Draw inscribed angle XZY.

3. Construct the line parallel to $\overleftrightarrow{ZY}$ through O.

   Label as R the point between X and Y where it intersects the circle.

4. Angle XOR is _ _ _ _ _ _ _ _ _ _ _ _ _ _ _ angle XZY.

   (a) smaller than   (c) larger than

   (b) congruent to

5. Compare angle XOR and angle XOY.

   Angle XOR is _ _ _ _ _ _ _ _ _ _ _ _ _ _ _ angle XOY.

   (a) one-third as large as   (c) congruent to

   (b) half as large as

6. Angle XZY is _ _ _ _ _ _ _ _ _ _ _ _ _ _ _ angle XOY.

   (a) one-third as large as   (c) congruent to

   (b) half as large as

<u>Review</u>

1. Construct a perpendicular to each line through the given point on the line.

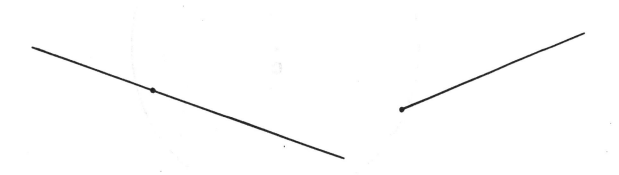

2. Check: Is the angle a right angle? _ _ _ _ _ _

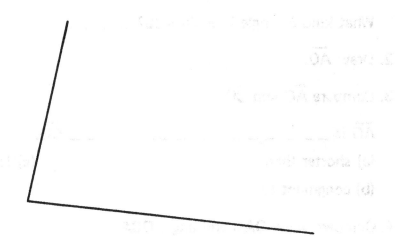

3. Compare the angles.

   Are they congruent? _ _ _ _ _

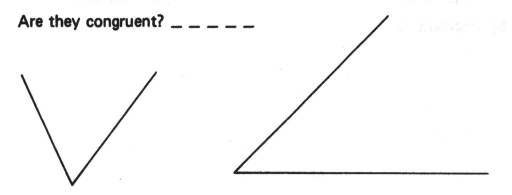

84

O is the center of the circle.

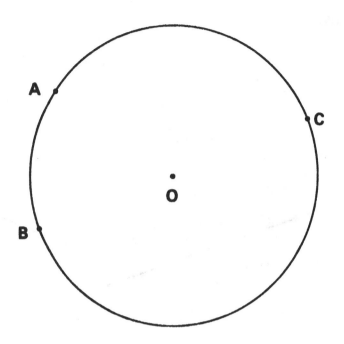

1. Draw angle ACB.

   What kind of angle is angle ACB? _ _ _ _ _

2. Draw $\overline{AO}$.

3. Compare $\overline{AO}$ and $\overline{CO}$.

   $\overline{AO}$ is _ _ _ _ _ _ _ _ _ _ _ _ _ _ _ $\overline{CO}$.
   (a) shorter than   (c) longer than
   (b) congruent to

4. Compare angle OAC and angle OCA.

   Angle OAC is _ _ _ _ _ _ _ _ _ _ _ _ _ _ angle OCA.
   (a) smaller than   (c) larger than
   (b) congruent to

Inscribed Angles

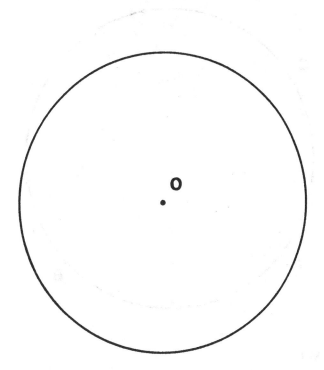

1. Draw a line through center O of the circle.

   Label as A and B the points where the line intersects the circle.

2. Choose any other point on the circle and label it C.

3. Draw angle ACB.

4. Check: Is angle ACB a right angle? _ _ _ _ _

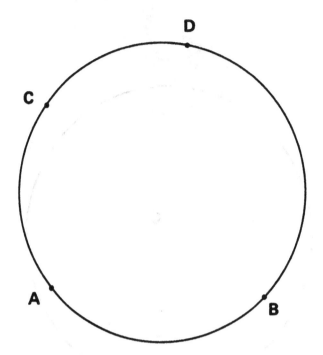

1. Draw angle ACB.

2. Draw angle ADB.

3. Compare angle ACB and angle ADB.

   Are they congruent? _ _ _ _ _

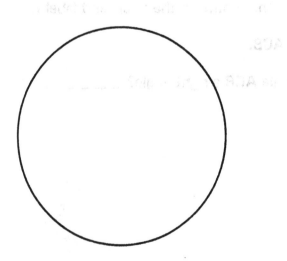

4. Choose 4 points on the circle and label them P, Q, R, S in that order.

5. Draw angle PQS and angle PRS.

6. Check: Is angle PQS congruent to angle PRS? _ _ _ _ _

<u>Practice Test</u>

1. Construct an angle which is congruent to the given angle.

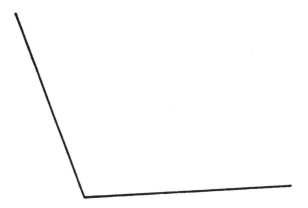

2. Triple the given segment.

3. Construct an angle which is double the given angle.

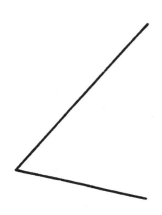

4. Divide segment $\overline{PQ}$ into five congruent parts.

5. Draw a segment which is 6/5 as long as segment $\overline{PQ}$.
   Label its endpoints X and Y.

6. Draw a segment which is 1-1/5 as long as segment $\overline{PQ}$.
   Label its endpoints V and W.

7. Compare segments $\overline{XY}$ and $\overline{VW}$.
   Are they congruent? _ _ _ _ _ _

8. Bisect angle ABC.

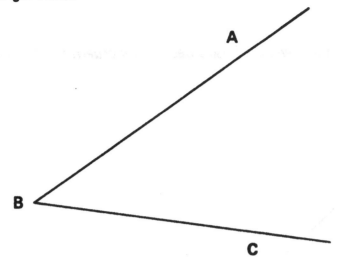

9. Construct an angle 5/2 as large as angle ABC.
   Label the angle RST.

10. Construct an angle 2-1/2 times as large as angle ABC.
    Label the angle XYZ.

11. Compare angle RST and angle XYZ.
    Are they congruent? _ _ _ _ _

Constructing a Triangle with Sides Congruent to the Sides of a Given Triangle

Problem: *Construct a triangle whose sides are congruent to the sides of a given triangle.*

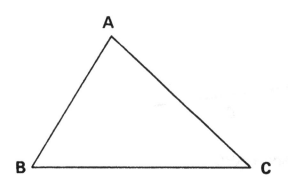

Solution:

1. Draw a line and on it lay off a segment congruent to segment $\overline{BC}$. Label the endpoints of the segment D and G.

2. Draw an arc with center D and radius congruent to $\overline{AB}$.

3. Draw an arc with center G and radius congruent to $\overline{AC}$. Make the arcs intersect.

4. Label the point of intersection O.

5. Draw triangle DOG.

6. The sides of triangle DOG are _ _ _ _ _ _ _ _ _ _ _ _ _ _ _ _ _ _ the sides of triangle ABC.

   (a) twice as long as   (c) congruent to

   (b) half as long as

7. Side $\overline{AB}$ corresponds to side _ _ _ _ _ .

   Side $\overline{AC}$ corresponds to side _ _ _ _ _ .

   Side $\overline{BC}$ corresponds to side _ _ _ _ _ .

1. Construct a triangle whose sides are congruent to the sides of triangle XYZ.

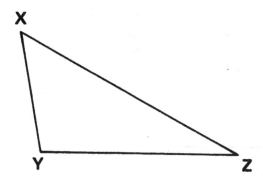

2. Label the triangle MNP.

3. Side $\overline{XY}$ corresponds to side _ _ _ _ _ .

   Side $\overline{XZ}$ corresponds to side _ _ _ _ _ .

   Side $\overline{YZ}$ corresponds to side _ _ _ _ _ .

4. Angle XZY corresponds to angle _ _ _ _ _ .

   Angle XYZ corresponds to angle _ _ _ _ _ .

   Angle YXZ corresponds to angle _ _ _ _ _ .

5. Check: Are the corresponding angles congruent? _ _ _ _ _

Problem: *Construct a triangle with the given segments as sides.*

A ———————————— B

C ——————————— D

E ——————— F

—————————————————

Solution:

1. On the line, lay off a segment congruent to $\overline{AB}$.
   Label the endpoints of the segment X and Y.

2. Draw an arc with center X and radius congruent to $\overline{CD}$.

3. Draw an arc with center Y and radius congruent to $\overline{EF}$.
   Make the arcs intersect.
   Label the intersection Z.

4. Draw $\overline{ZX}$ and $\overline{ZY}$.

5. Side $\overline{ZX}$ is congruent to segment _ _ _ _ _ .

   Side $\overline{ZY}$ is congruent to segment _ _ _ _ _ .

   Side $\overline{XY}$ is congruent to segment _ _ _ _ _ .

————————

——————————

———————————

6. Construct a triangle whose sides are congruent to the given segments.

Review

1. Bisect each segment.

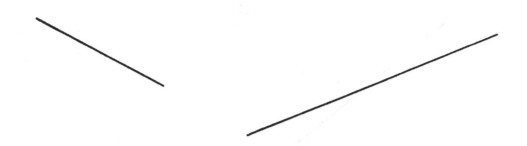

2. Construct a triangle whose sides are congruent to the given triangle.

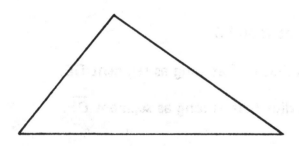

3. Compare the angles.

   Are they congruent? _ _ _ _ _

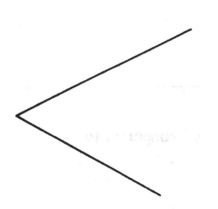

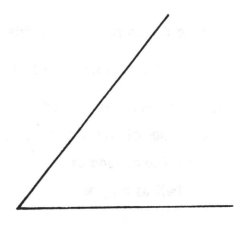

Constructing a Triangle with Sides Half those of a Given Triangle

Problem: *Construct a triangle whose sides are half as long as the sides of a given triangle.*

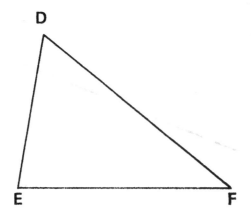

Solution:

1. Bisect each side of triangle DEF.

2. Draw a line and on it lay off a segment whose length is half as long as segment $\overline{EF}$.

3. Label the endpoints of the segment R and S.

4. Draw an arc with center R and radius half as long as segment $\overline{DE}$.

5. Draw an arc with center S and radius half as long as segment $\overline{DF}$. Make the arcs intersect.

6. Label as T the point where the arcs intersect.

7. Draw triangle RST.

8. Side $\overline{RS}$ corresponds to side _ _ _ _ _ .

   Side $\overline{RT}$ corresponds to side _ _ _ _ _ .

   Side $\overline{ST}$ corresponds to side _ _ _ _ _ .

9. The sides of triangle RST are _ _ _ _ _ _ _ _ _ _ _ _ _ _ _ the sides of triangle DEF.

   (a) twice as long as   (c) congruent to

   (b) half as long as

1. Construct a triangle whose sides are half as long as the sides of triangle ABC.

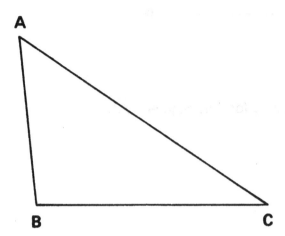

2. Label the triangle UVW.

3. Side $\overline{UV}$ corresponds to side _ _ _ _ _ .

   Side $\overline{UW}$ corresponds to side _ _ _ _ _ .

   Side $\overline{VW}$ corresponds to side _ _ _ _ _ .

4. Angle UVW corresponds to angle _ _ _ _ _ .

   Angle UWV corresponds to angle _ _ _ _ _ .

   Angle WUV corresponds to angle _ _ _ _ _ .

5. Compare the corresponding angles.

   Are they congruent? _ _ _ _ _

Review

A  B

1. Draw a segment which is twice as long as segment $\overline{AB}$.

2. Draw a segment which is three times as long as segment $\overline{AB}$.

3. Construct an equilateral triangle.

Constructing a Triangle with Sides Twice those of a Given Triangle

Problem: *Construct a triangle whose sides are twice as long as the sides of a given triangle.*

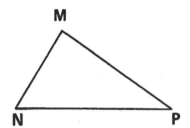

Solution:

1. On the line, lay off a segment twice as long as side $\overline{NP}$.
   Label its endpoints A and B.

2. Double segment $\overline{MN}$.

3. Draw an arc with center A and radius double segment $\overline{MN}$.

4. Double segment $\overline{MP}$.

5. Draw an arc with center B and radius double segment $\overline{MP}$.
   Make the arcs intersect.

6. Label the intersection C.

7. Side $\overline{AB}$ corresponds to side _ _ _ _ _ .

   Side $\overline{AC}$ corresponds to side _ _ _ _ _ .

   Side $\overline{BC}$ corresponds to side _ _ _ _ _ .

8. Angle ABC corresponds to angle _ _ _ _ _ .

   Angle CAB corresponds to angle _ _ _ _ _ .

   Angle ACB corresponds to angle _ _ _ _ _ .

1. Construct a triangle whose sides are twice as long as sides of triangle PQR.

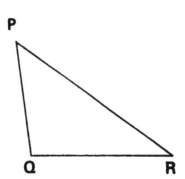

2. Label the triangle STV.

3. Angle PQR corresponds to angle _ _ _ _ _ .

   Angle PRQ corresponds to angle _ _ _ _ _ .

   Angle QPR corresponds to angle _ _ _ _ _ .

4. Compare the corresponding angles.

   The angles of triangle STV are _ _ _ _ _ _ _ _ _ _ _ _ _ _ _ _
   the corresponding angles of triangle PQR.

   (a) twice as large as   (c) congruent to
   (b) half as large as

6. Side $\overline{QR}$ corresponds to side _ _ _ _ _ .

   Side $\overline{PQ}$ corresponds to side _ _ _ _ _ .

   Side $\overline{PR}$ corresponds to side _ _ _ _ _ .

7. The sides of triangle STV are _ _ _ _ _ _ _ _ _ _ _ _ _ _ _
   the corresponding sides of triangle PQR.

   (a) half as long as   (c) congruent to
   (b) twice as long as

1. Construct a triangle whose sides are half as long as the sides of triangle FOG.

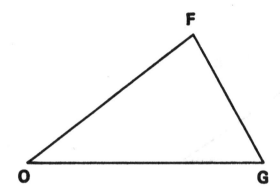

2. Construct a triangle whose sides are twice as long as the sides of triangle FOG.

Construct a triangle whose sides are three times as long as the sides of
the given triangle.

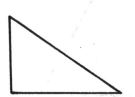

Constructing a Triangle with Sides One-third those of a Given Triangle

1. Construct a triangle whose sides are congruent to segments $\overline{AB}$, $\overline{CD}$, and $\overline{EF}$.

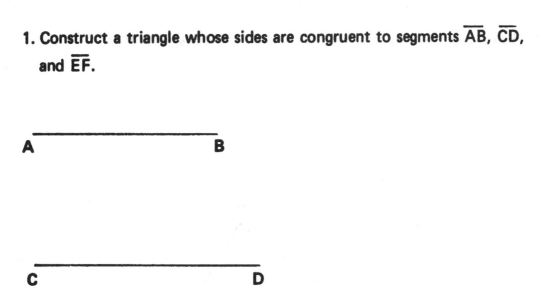

A   B

C   D

E   F

2. Divide segments $\overline{AB}$, $\overline{CD}$, and $\overline{EF}$ into three congruent parts.

3. Construct a triangle whose sides are 1/3 as long as segments $\overline{AB}$, $\overline{CD}$, and $\overline{EF}$.

1. Divide each side of triangle ABC into three congruent parts.

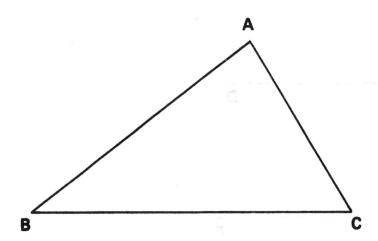

2. Construct a triangle whose sides are 2/3 as long as the sides of triangle ABC.

Review

1. Construct an angle congruent to angle ABC.

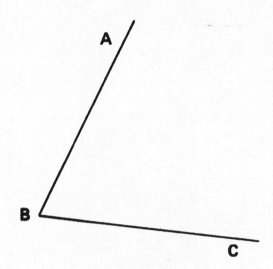

2. Construct an angle twice as large as angle ABC.

3. Construct the altitude from E to side $\overline{DF}$

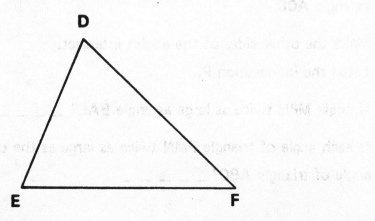

Do you think you can construct a triangle whose angles are twice as large as the angles of triangle ABC? _ _ _ _ _

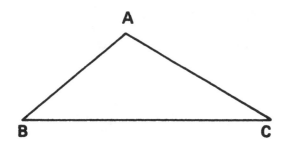

1. Compare segment $\overline{BC}$ and segment $\overline{MN}$.

   Are they congruent? _ _ _ _ _

2. Construct an angle at M, with $\overrightarrow{MN}$ as one side, which is twice as large as angle ABC.

3. Construct an angle at N, with $\overrightarrow{NM}$ as one side, which is twice as large as angle ACB.

4. Make the other sides of the angles intersect.
   Label the intersection P.

5. Is angle MPN twice as large as angle BAC? _ _ _ _ _

6. Is each angle of triangle PMN twice as large as the corresponding angle of triangle ABC? _ _ _ _ _

Similar Triangles

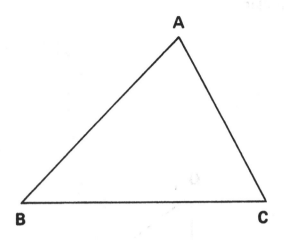

1. Bisect each side of triangle ABC.

2. Check: Is segment $\overline{XY}$ half as long as segment $\overline{BC}$? _ _ _ _ _

3. Construct an angle with vertex X and side $\overrightarrow{XY}$ which is congruent to angle ABC.

4. Construct an angle with vertex Y and side $\overrightarrow{YX}$ which is congruent to angle ACB.

5. Make the other sides of the angles intersect.
   Label the point of intersection Z.

6. Check: Is angle XZY congruent to angle BAC? _ _ _ _ _

7. Two triangles are <u>similar</u> if their corresponding angles are congruent.

   Are triangle ABC and triangle XYZ similar? _ _ _ _ _

   Why? _

8. Check: Is side $\overline{XZ}$ half as long as side $\overline{AB}$? _ _ _ _ _

   Check: Is side $\overline{YZ}$ half as long as side $\overline{AC}$? _ _ _ _ _

   Is each side of triangle XYZ half as long as the corresponding side of

   triangle ABC? _ _ _ _ _

Triangle ABC and triangle DEF are similar.

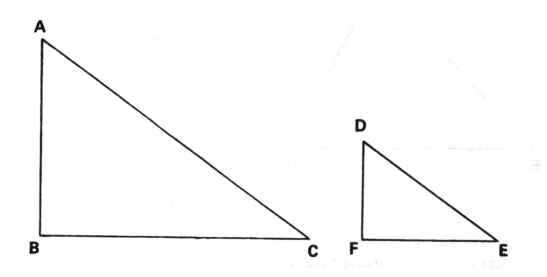

1. Side $\overline{AB}$ corresponds to side _ _ _ _ _ .

   Side $\overline{AC}$ corresponds to side _ _ _ _ _ .

   Side $\overline{BC}$ corresponds to side _ _ _ _ _ .

2. Check: Are the sides of triangle ABC double the corresponding sides

   of triangle DEF? _ _ _ _ _

3. Angle BAC corresponds to angle _ _ _ _ _ .

   Angle ABC corresponds to angle _ _ _ _ _ .

   Angle ACB corresponds to angle _ _ _ _ _ .

4. Check: Are the corresponding angles congruent? _ _ _ _ _

Triangle MNP is similar to triangle QRS.

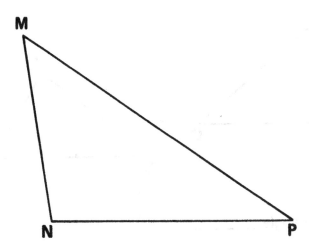
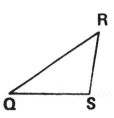

1. Side $\overline{MN}$ corresponds to side _ _ _ _ _ .

   Side $\overline{MP}$ corresponds to side _ _ _ _ _ .

   Side $\overline{NP}$ corresponds to side _ _ _ _ _ .

2. Check: Are the sides of triangle MNP triple the corresponding sides of triangle QRS? _ _ _ _ _

3. Angle MNP corresponds to angle _ _ _ _ _ .

   Angle MPN corresponds to angle _ _ _ _ _ .

   Angle MNP corresponds to angle _ _ _ _ _ .

4. Are the corresponding angles congruent? _ _ _ _ _

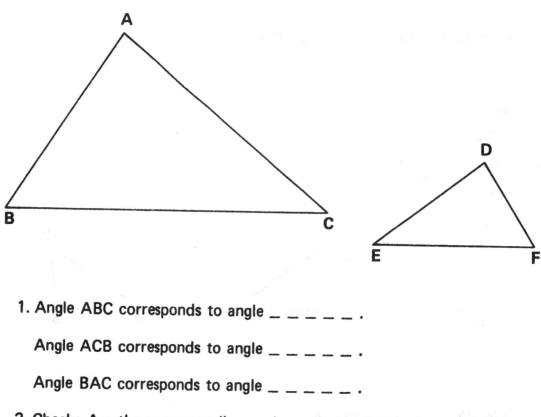

1. Angle ABC corresponds to angle _ _ _ _ _ .

   Angle ACB corresponds to angle _ _ _ _ _ .

   Angle BAC corresponds to angle _ _ _ _ _ .

2. Check: Are the corresponding angles congruent? _ _ _ _ _

3. Are the triangles similar? _ _ _ _ _

4. Check: Is side $\overline{AB}$ twice as long as side $\overline{DF}$? _ _ _ _ _

   Check: Is side $\overline{BC}$ twice as long as side $\overline{EF}$? _ _ _ _ _

   Check: Is side $\overline{AC}$ twice as long as side $\overline{DE}$? _ _ _ _ _

5. Check: Are the corresponding angles of the triangles below congruent? _ _ _ _ _

   Check: Are the triangles below similar? _ _ _ _ _

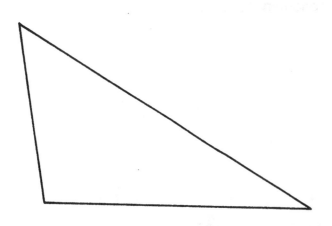

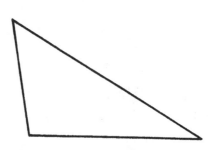

Problem: *Reproduce triangle ROM.*

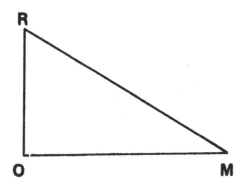

Solution:

1. On the given line, lay off a segment congruent to $\overline{OM}$.
   Label its endpoints L and K.

2. Reproduce angle ROM at L with $\overrightarrow{LK}$ as one side.

3. Reproduce angle RMO at K with $\overrightarrow{KL}$ as one side.

4. Make the other sides of the angle intersect.
   Label the intersection G.

5. Angle RMO is congruent to angle _ _ _ _ _ .

   Angle ROM is congruent to angle _ _ _ _ _ .

   Angle ORM is congruent to angle _ _ _ _ _ .

6. Are triangle ROM and triangle GLK congruent? _ _ _ _ _

   Are triangle ROM and triangle GLK similar? _ _ _ _ _

   Why? _

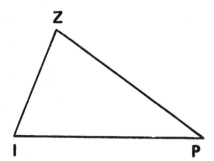

1. Draw a line and on it lay off a segment twice as long as segment $\overline{IP}$. Label its endpoints R and S.

2. Reproduce angle ZIP at R with $\overrightarrow{RS}$ as one side.

3. Reproduce angle ZPI at S with $\overrightarrow{SR}$ as one side.

4. Make the other sides of the angles intersect. Label the intersection T.

5. Angle TRS is congruent to angle _ _ _ _ _ _ .

   Angle TSR is congruent to angle _ _ _ _ _ _ .

   Angle RTS is congruent to angle _ _ _ _ _ _ .

6. Is triangle RST similar to triangle ZIP? _ _ _ _ _

   Why? _

7. Check: Is $\overline{TR}$ twice as long as $\overline{ZI}$? _ _ _ _ _

   Check: Is $\overline{TS}$ twice as long as $\overline{ZP}$? _ _ _ _ _

8. Triangle ZIP and triangle TRS are _ _ _ _ _ _ _ _ _ _ _ _ _ _ _ _ .

Review

1. Divide segment $\overline{ST}$ into five congruent parts.

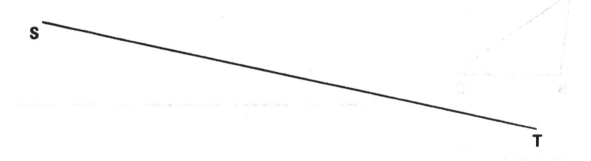

2. Each segment is _ _ _ _ _ _ _ _ _ _ _ _ _ _ _ _ _ of segment $\overline{ST}$.

   (a) one-half   (b) one-third   (c) one-fifth

3. Construct an equilateral triangle with each side 2/5 as long as segment $\overline{ST}$.

Constructing a Triangle Similar to a Given Triangle

Problem: *Construct a triangle similar to a given triangle whose sides are three times as long as the sides of the given triangle.*

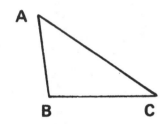

Solution:

1. On the given line, lay off a segment which is three times as long as segment $\overline{BC}$.
   Label its endpoints D and E.

2. Reproduce angle ABC at D, with $\overrightarrow{DE}$ as one side.

3. Reproduce angle ACB at E with $\overrightarrow{ED}$ as one side.
   Make the other sides of the angles intersect.

4. Label the point of intersection F.

5. Check: Is angle BAC congruent to angle DFE? _ _ _ _ _

6. Is triangle DEF similar to triangle ABC? _ _ _ _ _

7. Check: Is side $\overline{DF}$ three times as long as side $\overline{AB}$? _ _ _ _ .

   Check: Is side $\overline{EF}$ three times as long as side $\overline{AC}$? _ _ _ _ _

8. Triangle ABC and triangle FDE are _ _ _ _ _ _ _ _ _ _ _ _ _ _ _ .

1. Construct a triangle similar to triangle ABC whose sides are half as long as the sides of triangle ABC.

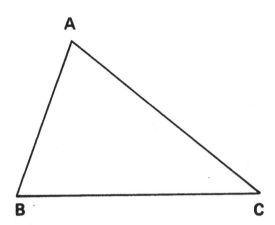

2. Construct a triangle similar to triangle ABC whose sides are twice as long as the sides of triangle ABC.

Problem: *Construct a triangle similar to a given triangle whose sides are 1-1/2 times as long as the sides of the given triangle.*

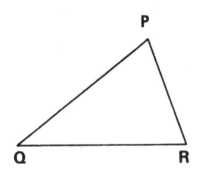

Solution:

1. Bisect side $\overline{QR}$.

2. On the given line, lay off a segment 1-1/2 times as long as side $\overline{QR}$. Label its endpoints A and B.

3. Reproduce angle PQR at A, with $\overrightarrow{AB}$ as one side.

4. Reproduce angle PRQ at B, with $\overrightarrow{BA}$ as one side. Make the other sides of the angles intersect.

5. Label the intersection C.

6. Are the triangles similar? _ _ _ _ _

7. Bisect sides $\overline{PQ}$ and $\overline{PR}$.

8. Check: Is side $\overline{CA}$ 1-1/2 times as long as side $\overline{PQ}$? _ _ _ _ _ _

   Check: Is side $\overline{CB}$ 1-1/2 times as long as side $\overline{PR}$? _ _ _ _ _ _

9. Triangle PQR and triangle CAB are _ _ _ _ _ _ _ _ _ _ _ _ _ _ _ _ .

Problem: *Construct a triangle similar to a given triangle with a given side.*

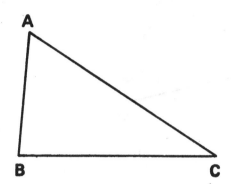

S ———————————————————————— U

Solution:

1. Reproduce angle ABC at S with $\overrightarrow{SU}$ as one side.

2. Reproduce angle ACB at U with $\overrightarrow{US}$ as one side.

3. Make the other sides of the angles intersect.
   Label the intersection V.

4. Is triangle SVU similar to triangle BAC? _ _ _ _ _

5. Construct a triangle similar to triangle XYZ, with the given segment
   as the side corresponding to side $\overline{YZ}$.

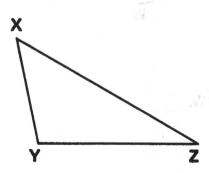

<u>Review</u>

1. Divide segment $\overline{PQ}$ into four congruent parts.

2. Draw a segment which is 3/4 as long as segment $\overline{PQ}$.

3. Divide segment $\overline{VW}$ into three congruent parts.

4. Draw a segment which is 1-1/3 as long as $\overline{VW}$.

1. Divide side $\overline{BC}$ into four congruent parts.

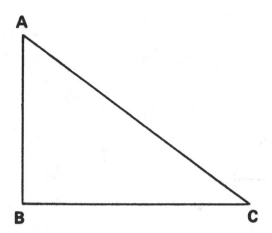

2. Construct a triangle similar to triangle ABC whose sides are 3/4 as long as the sides of triangle ABC.

3. Construct a triangle similar to triangle ABC whose sides are 3/2 as long as the sides of triangle ABC.

1. Construct a triangle similar to triangle ABC whose sides are 2/3 as long as the sides of triangle ABC.

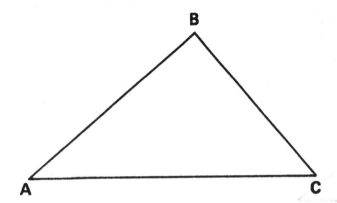

2. Label the triangle DEF.

3. Construct a triangle similar to triangle DEF whose sides are 3/2 as long as the sides of triangle DEF.

4. Label the triangle MNP.

5. Is triangle MNP similar to triangle ABC? _ _ _ _ _

6. Is triangle MNP congruent to triangle ABC? _ _ _ _ _

1. Construct an equilateral triangle with segment $\overline{RS}$ as one side.

2. Bisect the given angle.

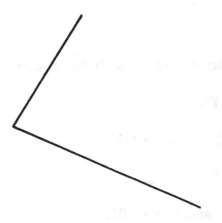

3. Construct a triangle similar to triangle DEF with segment $\overline{AB}$ as the side corresponding to side $\overline{DE}$.

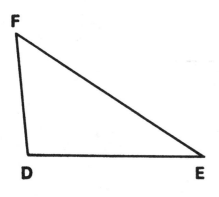

Similar Equilateral and Isosceles Triangles

1. Check: Is triangle PQR an equilateral triangle? _ _ _ _ _

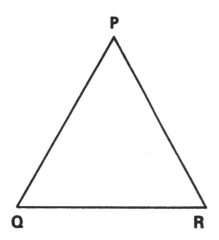

A _____ B

2. Construct an equilateral triangle with $\overline{AB}$ as one side.

3. Label the triangle ABC.

4. Compare angle ABC and angle PQR.

   Are they congruent? _ _ _ _ _

5. Compare angle ACB and angle PRQ.

   Are they congruent? _ _ _ _ _

6. Compare angle BAC and angle QPR.

   Are they congruent? _ _ _ _ _

7. Are the equilateral triangles similar? _ _ _ _ _

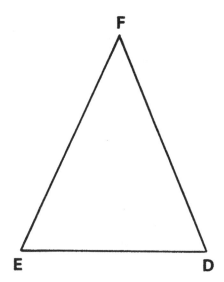

1. Compare sides $\overline{FE}$ and $\overline{DF}$.

   Are they congruent? _ _ _ _ _

2. At X construct an angle congruent to angle EFD.

3. Draw an arc with center X intersecting both sides of the angle.
   Label the points of intersection Y and Z.

4. Draw $\overline{YZ}$.

5. Check: Is angle XYZ congruent to angle FED? _ _ _ _ _

   Check: Is angle XZY congruent to angle FDE? _ _ _ _ _ _

6. Is triangle XYZ similar to triangle EFD? _ _ _ _ _

Similar Polygons

1. Bisect each side of triangle PQR.

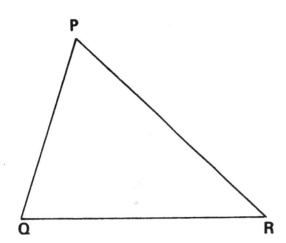

2. On the given line, lay off a segment half as long as $\overline{QR}$.

3. Label its endpoints A and B.

4. Construct an angle congruent to angle PQR at A with $\overrightarrow{AB}$ as one side.

5. On the other side of the angle, lay off a segment half as long as $\overline{QP}$ with A as one endpoint.
   Label its other endpoint C.

6. Draw triangle ABC.

7. Check: Is angle CBA congruent to angle PRQ? _ _ _ _ _

   Check: Is angle ACB congruent to angle QPR? _ _ _ _ _

8. Check: Is side CB half as long as side PR? _ _ _ _ _

9. Are the triangles similar? _ _ _ _ _

Problem: *Construct a quadrilateral similar to a given quadrilateral whose sides are half as long as the sides of the given quadrilateral.*

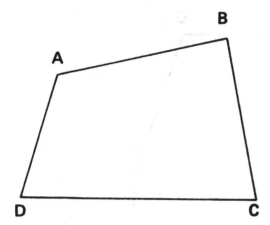

Solution:

1. Bisect each side of quadrilateral ABCD.

2. On the given line, lay off a segment half as long as $\overline{DC}$.
   Label its endpoints F and G.

3. At F construct an angle congruent to angle ADC with $\overrightarrow{FG}$ as one side.

4. Draw an arc with center F and radius half as long as side $\overline{AD}$ which intersects the other side of the angle.
   Label the point of intersection J.

5. At G construct an angle congruent to angle BCD with $\overrightarrow{GF}$ as one side.

6. Draw an arc with center G and radius half as long as side $\overline{BC}$ which intersects the other side of the angle.
   Label the intersection H.

7. Draw $\overline{JH}$.

8. Quadrilateral FGHJ is similar to quadrilateral ABCD.

Construct a quadrilateral similar to quadrilateral PQRS whose sides are twice as long as the sides of PQRS.

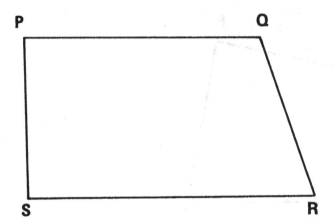

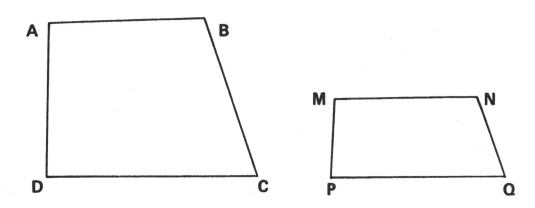

1. Angle DAB corresponds to angle _ _ _ _ _ .

   Angle ABC corresponds to angle _ _ _ _ _ .

   Angle ADC corresponds to angle _ _ _ _ _ .

   Angle BCD corresponds to angle _ _ _ _ _ .

2. Check: Are the corresponding angles congruent? _ _ _ _ _

3. Bisect the sides of ABCD.

4. Check: Is side $\overline{MP}$ half as long as side $\overline{AD}$? _ _ _ _ _

   Check: Is side $\overline{PQ}$ half as long as side $\overline{DC}$? _ _ _ _ _

   Check: Is side $\overline{NQ}$ half as long as side $\overline{BC}$? _ _ _ _ _

   Check: Is side $\overline{MN}$ half as long as side $\overline{AB}$? _ _ _ _ _

5. Do you think the quadrilaterals are similar? _ _ _ _ _

   Why? _

6. Two quadrilaterals are _ _ _ _ _ _ _ _ _ _ _ _ _ _ _ _ _ similar
   if their corresponding angles are congruent.

   (a) always   (b) sometimes   (c) never

Review

1. Construct a triangle with sides half as long as the sides of the given triangle.

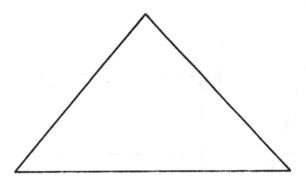

2. Construct a line parallel to $\overleftrightarrow{AB}$ through point C.

C.

A   B

3. Construct a line perpendicular to $\overleftrightarrow{AB}$ through point C.

The Triangle Formed by the Midpoints of the Sides of a Given Triangle

1. Bisect side $\overline{AB}$ of triangle ABC.

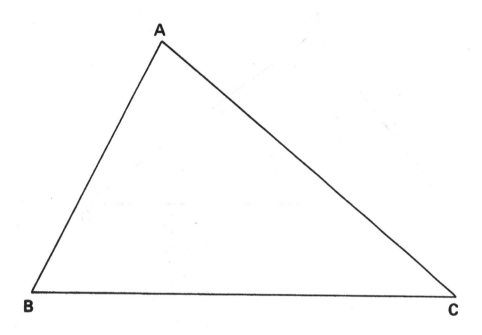

2. Label as T the midpoint of side $\overline{AB}$.

3. Construct the line parallel to $\overleftrightarrow{BC}$ through T.

4. Label as V the point where the parallel intersects $\overline{AC}$.

5. Construct the line parallel to $\overleftrightarrow{AB}$ through V.

6. Label as W the point where the parallel intersects $\overline{BC}$.

7. Draw triangle TVW.

8. Angle TWV is congruent to angle _ _ _ _ _ .

   Angle WTV is congruent to angle _ _ _ _ _ .

   Angle TVW is congruent to angle _ _ _ _ _ .

9. Is triangle TVW similar to triangle ABC? _ _ _ _ _

   Why? _

1. Bisect the sides of triangle PQR.

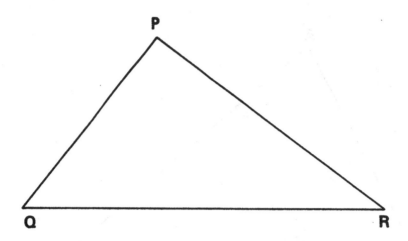

2. Label the midpoints M, N, and O.

3. Draw triangle MNO.

4. Triangles MNO and PQR are similar.

   Side $\overline{MN}$ corresponds to side _ _ _ _ _ .

   Side $\overline{MO}$ corresponds to side _ _ _ _ _ .

   Side $\overline{NO}$ corresponds to side _ _ _ _ _ .

5. Each side of triangle MNO is _ _ _ _ _ _ _ _ _ _ _ _ _ _ _ _ _
   the corresponding side of triangle PQR.

   (a) half as long as   (c) twice as long as

   (b) congruent to   (d) three times as long as

1. Bisect side $\overline{QR}$.

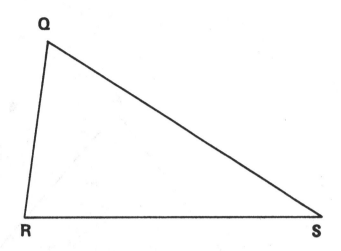

2. Label as M the midpoint of $\overline{QR}$.

3. Construct a line parallel to $\overleftrightarrow{RS}$ through M.

4. Label as N the point where the parallel intersects side $\overline{QS}$.

5. Draw $\overline{MN}$.

6. Is angle QMN congruent to angle QRS? _ _ _ _ _

   Is angle QNM congruent to angle QSR? _ _ _ _ _

7. Check: Is $\overline{QN}$ half as long as $\overline{QS}$? _ _ _ _ _

   Check: Is $\overline{MN}$ half as long as $\overline{RS}$? _ _ _ _ _

8. Is triangle QMN similar to triangle QRS? _ _ _ _ _

1. Construct the altitude from A to side $\overline{BC}$.

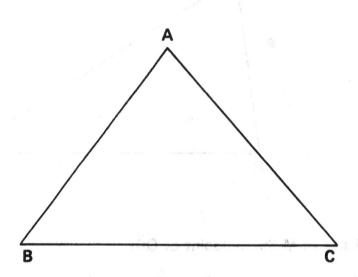

2. Label as P the point where it intersects side $\overline{BC}$.

3. Construct the altitude from C to side $\overline{AB}$.

4. Label as Q the point where it intersects side $\overline{AB}$.

5. Construct the altitude from B to side $\overline{AC}$.

6. Label as R the point where it intersects side $\overline{AC}$.

7. Draw triangle PQR.

8. Is triangle PQR similar to triangle ABC? _ _ _ _ _ _

Similar Triangles in a Circle

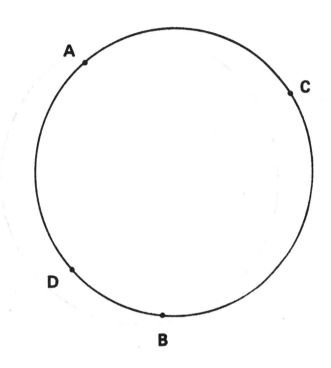

1. Draw $\overline{AB}$.

2. Draw $\overline{CD}$.

   Label the point of intersection E.

3. Draw triangle AED and triangle CEB.

4. Do you think triangle AED and triangle CEB are similar? _ _ _ _ _

5. Angle AED corresponds to angle _ _ _ _ _ .

   Angle DAE corresponds to angle _ _ _ _ _ .

   Angle ADE corresponds to angle _ _ _ _ _ .

6. Check: Are the corresponding angles congruent? _ _ _ _ _

7. Are the triangles similar? _ _ _ _ _

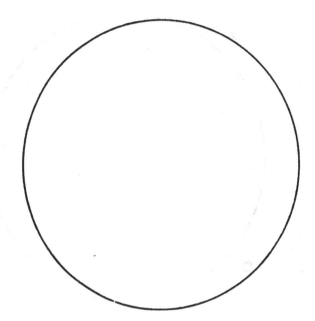

1. Draw a line through the circle.

   Label as P and Q the points where it intersects the circle.

2. Draw another line through the circle which intersects line $\overleftrightarrow{PQ}$ inside the circle.

3. Label as T the point where the lines intersect.

   Label as R and S the points where the line intersects the circle.

4. Draw $\overline{RP}$ and $\overline{SQ}$.

5. Do you think triangle RPT is similar to triangle SQT? _ _ _ _ _

6. Angle RTP corresponds to angle _ _ _ _ _ .

   Angle TRP corresponds to angle _ _ _ _ _ .

   Angle RPT corresponds to angle _ _ _ _ _ .

7. Check: Are the corresponding angles congruent? _ _ _ _ _

8. Are the triangles similar? _ _ _ _ _

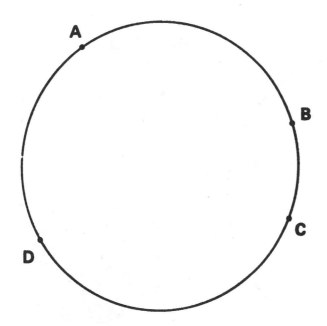

1. Draw a line through A and B.

2. Draw a line through C and D.
   Make the lines intersect.

3. Label the point of intersection E.

4. Draw $\overline{AC}$ and $\overline{BD}$.

5. Do you think triangle ACE and triangle DBE are similar? _ _ _ _ _

6. Angle CAE corresponds to angle _ _ _ _ _ .

   Angle ACE corresponds to angle _ _ _ _ _ .

   Angle AEC corresponds to angle _ _ _ _ _ .

7. Check: Are the corresponding angles congruent? _ _ _ _ _

8. Are the triangles similar? _ _ _ _ _

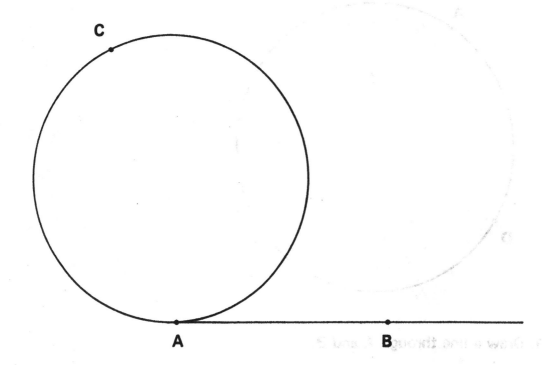

1. Draw line $\overleftrightarrow{BC}$.

2. Label as D the other point where the line intersects the circle.

3. Draw $\overline{AC}$ and $\overline{AD}$.

4. Do you think triangle ACB is similar to triangle ADB? _ _ _ _ _

5. Angle ACB corresponds to angle _ _ _ _ _

   Angle CAB corresponds to angle _ _ _ _ _

   Angle ABC corresponds to angle _ _ _ _ _

6. Check: Are the corresponding angles congruent? _ _ _ _ _

7. Are the triangles similar? _ _ _ _ _

Practice Test

1. Divide segment $\overline{AB}$ into three congruent parts.

2. Bisect each angle of the triangle.

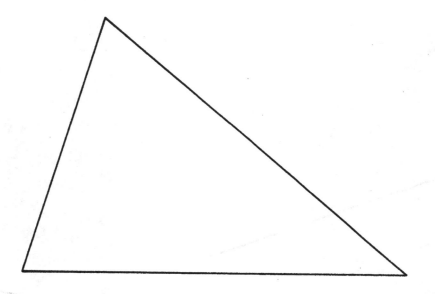

3. Construct a triangle similar to triangle ABC with $\overline{MN}$ as the side corresponding to side $\overline{BC}$.

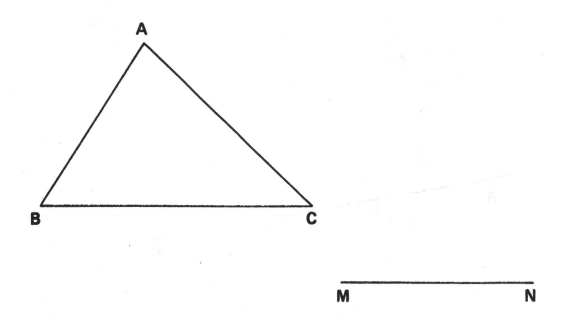

4. Construct a perpendicular to each line through the given point.

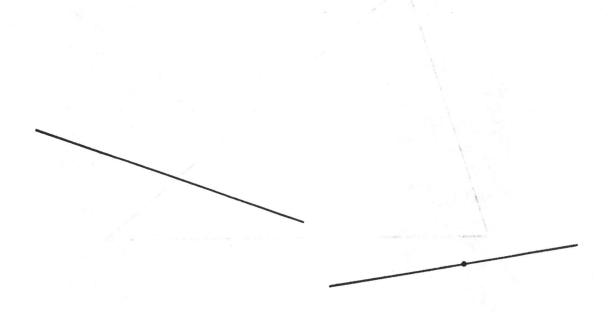

5. Construct a line parallel to the given line.

6. Construct a quadrilateral similar to the given quadrilateral whose sides are half as long as the sides of the given quadrilateral.

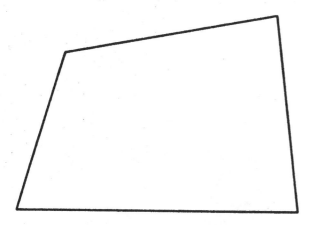

7. Construct the altitude from A to side $\overline{BC}$.

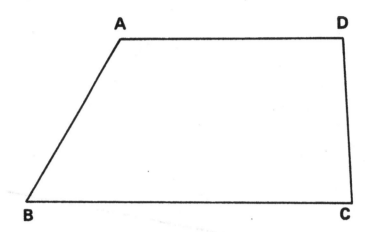

8. Construct a triangle whose sides are twice as long as the sides of triangle MNP.

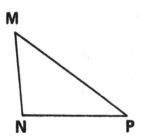

KEY TO GEOMETRY, Books 1-8
INDEX OF UNDERLINED TERMS

(Numbers indicate book and page.)